Agricultural Engineering

Agricultural Engineering

Maximus Stanley

Larsen & Keller
www.larsen-keller.com

Agricultural Engineering
Maximus Stanley
ISBN: 978-1-64172-344-2 (Hardback)

Larsen & Keller
Published by Larsen and Keller Education,
5 Penn Plaza,
19th Floor,
New York, NY 10001, USA

Cataloging-in-Publication Data

Agricultural engineering / Maximus Stanley.
p. cm.
Includes bibliographical references and index.
ISBN 978-1-64172-344-2
1. Agricultural engineering. 2. Engineering. 3. Agriculture.
4. Farm equipment. I. Stanley, Maximus.
S675 .A37 2020
630--dc23

Printed and bound in China.

For more information regarding Larsen and Keller Education and its products, please visit the publisher's website www.larsen-keller.com

TABLE OF CONTENTS

PREFACE

It is with great pleasure that I present this book. It has been carefully written after numerous discussions with my peers and other practitioners of the field. I would like to take this opportunity to thank my family and friends who have been extremely supporting at every step in my life.

The discipline of engineering that deals with agricultural production and processing is referred to as agricultural engineering. It is an interdisciplinary field that combines various other disciplines such as mechanical, civil, chemical and electrical engineering. Agricultural engineering aims to improve the effectiveness and sustainability of agricultural practices. It encompasses various specialties such as the designing of agricultural machinery, equipment and agricultural structures, water management for irrigation and livestock production. It is also concerned with soil management and conservation. Principles from this field are applied in a variety of areas such as flood water control systems, agricultural product processing, irrigation and devising dairy effluent schemes. The topics included in this book on agricultural engineering are of utmost significance and bound to provide incredible insights to readers. It covers in detail some existent theories and innovative concepts revolving around this field. This book is appropriate for students seeking detailed information in this area as well as for experts.

The chapters below are organized to facilitate a comprehensive understanding of the subject:

Chapter - Introduction

The branch of engineering which focuses on improving the efficacy and sustainability of agricultural processing and production is referred to as agricultural engineering. This is an introductory chapter which will provide a brief introduction to all the significant aspects of agricultural engineering such as agronomy and precision agriculture.

Chapter - Agroecology

The study of the ecological processes that are applied to the agricultural production systems is known as agroecology. Some of its focus areas are agroecological restoration, organic farming, carbon farming and regenerative agriculture. This chapter has been carefully written to provide an easy understanding of these aspects of agroecology.

Chapter - Agricultural Biotechnology

The field of agricultural science which is concerned with the utilization of scientific tools and techniques such as genetic engineering and molecular markers is known as agricultural biotechnology. Some of the crop modification techniques which are used within this field are traditional breeding, mutation breeding, protoplast fusion, transgenics, etc. This chapter closely examines these techniques of agricultural biotechnology.

Chapter – Agricultural Technology

The application of techniques for controlling the growth and harvesting of vegetable products is termed as agricultural technology. Some of the machines which fall under agricultural technology are agricultural robots, rollers, subsoilers, cultivators, etc. The topics elaborated in this chapter will help in gaining a better perspective about these agricultural technologies.

Chapter – Dairy Farming Technology

The class of agriculture which deals with the long-term production of milk is referred to as dairy farming. Some of the modern techniques and technologies used for dairy farming are artificial insemination, grain crimping, automatic milking, etc. The diverse applications of these dairy farming technologies have been thoroughly discussed in this chapter.

Maximus Stanley

Introduction 1

- Agricultural Engineering
- Precision Agriculture
- Agronomy

The branch of engineering which focuses on improving the efficacy and sustainability of agricultural processing and production is referred to as agricultural engineering. This is an introductory chapter which will provide a brief introduction to all the significant aspects of agricultural engineering such as agronomy and precision agriculture.

Agricultural Engineering

Agricultural Engineering relates to the science and technology of producing food grains. It applies principles of engineering science and technology to production and processing in the field of agriculture.

Plant Breeding

Plant breeding is a method of altering the genetic pattern of plants to increase their value and utility for human welfare. It is a purposeful manipulation of plants to create desired plant types that are better suited for cultivation, give better yield and are disease resistant. Plant breeding is done for the following objectives:

- Increase the crop yield;
- Improve the quality of the crop;
- Increase tolerance to environmental conditions like salinity, extreme temperatures and drought;
- Develop a resistance to pathogens;
- Increase tolerance to the insect pest.

Steps for Different Plant Breeding Methods

The main steps of the plant breeding program are as follows:

1. Collection of Variability: Wild varieties species and relatives of the cultivated species having desired traits should be collected and preserved. The entire collection having all the diverse alleles for all genes in a given crop is called *germplasm collection*. Germplasm conservation can be done following ways:

- In situ conservation: It can be done with the help of forests and Natural Reserves.
- Ex situ conservation: It is done through botanical gardens, seed banks.

2. Evaluation and Selection of Parents: The germplasm collected is evaluated to identify the plants with desirable characters. It is made sure that only the pure lines are selected. The selected plants are multiplied and used in the process of hybridization.

3. Hybridization: The Pollen Grain from one desired parent plant selected as a male parent is collected and dusted over another plant which is considered as the female parent.

4. Selection and Testing of Superior Recombinants: Progeny obtained after crossing are evaluated for the desired combination of characters. These are self-pollinated for several generations till there is a state of uniformity so that the characters will not segregate further.

5. Testing Release and Commercialization of New Cultivars: The selected plants are evaluated by growing the plants in an experimental field and the performance is recorded. This is done for at least 3 growing Seasons at different locations in the country.

Plant Breeding for Disease Resistance

Some of the diseases caused by plants are:

Fungi	Brown rust of wheat, Red rot of sugarcane, Late blight of potato.
Bacteria	Black rot of crucifers.
Viruses	Tobacco mosaic turnip music.

The basic objective of breeding for disease resistance is to develop inherent quality in the plant to prevent the pathogen from causing the disease. Such varieties of plants are called resistant plants. The basic method is the same as normal hybridization. For hybridization resistant plant should be available for breeding.

Some diseases resistant plants developed are:

Crop	Variety	Resistant to Disease
Wheat	Himgiri	Leaf & stipe rust
Brassica	Pusa Swarnim	White rust
Cauliflower	Pusa Shubra	Black rot and curl blight
Chilli	Pusa Sadabahar	Chilli mosaic virus

Mutation Breeding

If resistant variety is not available, the resistance can be developed by inducing mutations in the plant through diverse means and then by screening the plant material for resistance.

Mutation is changed in the base sequence of the genes, induced by certain Chemicals or radiations. The resistant plants developed can be multiplied directly or can be used in other breeding experiments.

Plant Breeding for Developing a Resistance to Insect Pest

Resistance can be developed by following ways:

- Development of morphological characters like hairy leaves in cotton and wheat develop vector resistance from jassids beetle.
- Solid stem in wheat lead to resistance from stem borers.
- Biochemical characters provide resistance to insects and pests. For example, the high aspartic acid and low nitrogen and sugar content in maize leads to resistance to maize stem borers.
- Smooth leaves and nectarless cotton develop resistance from bollworms.

Some pest-resistant varieties are:

Crop	Variety	Insect pest
Brassica	Pusa Gaurav	Aphids
Flat bean	Pusa sem	Jassids, aphids
Okra	Pusa Sawni	Shoot and fruit bores

Plant Breeding for Improved Food Quality

Biofortification is a method in which crops are bred for higher levels of vitamins, minerals, and fats. Due to this problem of malnutrition can be overcome. Following objectives were considered for the breeding program:

- Protein content and quality;
- Oil content and quality;
- Vitamin content;
- Micronutrient content and quality.

Some examples of biofortification:

- Fortified Maize having twice the amount of amino acids lysine and tryptophan;
- Atlas 66 wheat has a high protein content;
- Iron-fortified rice having 5 times more iron;
- Vegetable crops like carrot and spinach have more vitamins and minerals;
- Vitamin C enriched bitter gourd.

Precision Agriculture

Precision agriculture (PA) is an approach to farm management that uses information technology (IT) to ensure that the crops and soil receive exactly what they need for optimum health and productivity. The goal of PA is to ensure profitability, sustainability and protection of the environment. PA is also known as satellite agriculture, as-needed farming and site-specific crop management (SSCM).

Precision agriculture relies upon specialized equipment, software and IT services. The approach includes accessing real-time data about the conditions of the crops, soil and ambient air, along with other relevant information such as hyper-local weather predictions, labor costs and equipment availability. Predictive analytics software uses the data to provide farmers with guidance about crop rotation, optimal planting times, harvesting times and soil management.

Sensors in fields measure the moisture content and temperature of the soil and surrounding air. Satellites and robotic drones provide farmers with real-time images of individual plants. Information from those images can be processed and integrated with sensor and other data to yield guidance for immediate and future decisions, such as precisely what fields to water and when or where to plant a particular crop.

Agricultural control centers integrate sensor data and imaging input with other data, providing farmers with the ability to identify fields that require treatment and determine the optimum amount of water, fertilizers and pesticides to apply. This helps the farmer avoid wasting resources and prevent run-off, ensuring that the soil has just the right amount of additives for optimum health, while also reducing costs and controlling the farm's environmental impact.

In the past, precision agriculture was limited to larger operations which could support the IT infrastructure and other technology resources required to fully implement and benefit from the benefits of precision agriculture. Today, however, mobile apps, smart sensors, drones and cloud computing makes precision agriculture possible for farming cooperatives and even small family farms.

Agronomy

Agronmy is the branch of agricultural science that deals with methods which provide favorable environment to the crop for higher productively.

Boundaries and Scale

Crop management, and its scientific study agronomy, are part of a system that comprises the physical elements of the climate, soil and land, the biological constituents of the vegetation and soil, the economic opportunities and constraints of markets, sales and profit, and the social circumstances and preferences of those who work the land.

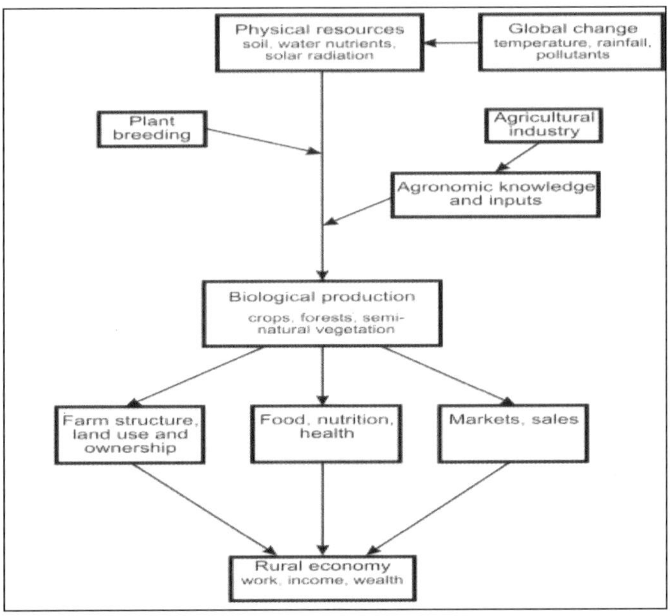

Flow diagram of physical, biological, economic and social dimensions of agronomy.

Scope of Agronomy

Agronomy is a dynamic discipline with the advancement of knowledge and better understanding of planet, environment and agriculture. Agronomy science becomes imperative in Agriculture in the following areas:

- Identification of proper season for cultivation of wide range of crops is needed which could be made possible only by Agronomy science.
- Proper methods of cultivation are needed to reduce the cost of cultivation and maximize the yield and economic returns.
- Availability and application of chemical fertilizers has necessitated the generation of knowledge to reduce the ill-effects due to excess application and yield losses due to the unscientific manner of application.
- Availability of herbicides for control of weeds has led to development for a vast knowledge about selectivity, time & method of its application.
- Water management practices play grater role in present day crisis of water demand and Agronomy science answer to the questions 'how much to apply?' and 'when to apply?'
- Intensive cropping is the need of the day and proper time and space intensification not only increase the production but also reduces the environmental hazards.
- New technology to overcome the effect of moisture stress under dry land condition is explored by Agronomy and future agriculture is depends on dry land agriculture.
- Packages of practices to explore full potential of new varieties of crops are the most important aspects in crop production which could be made possible only by Agronomy science.
- Keeping farm implements in good shape and utilizing efficient manner to nullify the present day labour crisis is further broadening the scope of agronomy.
- Maintaining the ecological balance through efficient management of crops, livestock and their feedings in a rational manner is possible only by knowing agronomic principles.
- Care and disposal of farm and animal products like milk and eggs and proper maintenance of accounts of all transactions concerning farm business is governing principles of agronomy.

Relation of Agronomy to other Sciences

Agronomy is a main branch of Agriculture. It is synthesis of several disciplines like soil science, Agricultural chemistry, crop physiology, plant ecology, biochemistry and economics.

The Soil Science helps the agronomist to thoroughly understand the soil physical, chemical and biological properties to effect modification of the soil environment.

The Agricultural Chemistry help the agronomist to understand the chemical composition and changes involved in the production, protection, and use of crops and livestock.

The crop physiology helps to understand the basic life process of crops to understand functioning of each parts of plant to determine their input requirement like nutrients etc.

The plant ecology helps us to understand the associated environment in which the crops grown like the influence of weather (Temperature, Rainfall etc).

The biochemistry shows the way in which biochemical process takes place in crops which helps to understand critical requirements to favourably activate this process.

The economics paves the way for profit and loss analysis in farming.

Agroecology 2

- **Agroecological Restoration**
- **Fertilizer Tree**
- **Organic Farming**
- **Bioeffector**
- **Good Agricultural Practice**
- **Regenerative Agriculture**
- **Carbon Farming**

The study of the ecological processes that are applied to the agricultural production systems is known as agroecology. Some of its focus areas are agroecological restoration, organic farming, carbon farming and regenerative agriculture. This chapter has been carefully written to provide an easy understanding of these aspects of agroecology.

Agroecology is holistic study of agro-ecosystems, including all environmental and human elements that focuses on the form, dynamics and functions of their inter- and intra-relationships. It can also be defined as an ecological approach to agriculture that views agricultural areas as ecosystems and is concerned with the ecological impact of agricultural practices. Agroecology allows deeper scientific understanding of agriculture as it applies ecological principles to systems of food production taking into account relationships between different components of the agroecosystem including the human community. It teaches us to be in tune with nature while producing a diversity of healthy, nutritive and delicious foods, using sources of nature. Agroecology, in essence, is the philosophy of relishing all edibles that nature produces and, at the same time, of nurturing nature to allow it blossom with its biodiversities.

Agroecology also delivers the social benefits associated with poverty reduction and community empowerment. At one hand where it reduces the possibility of environmental degradation, on other hand it is capable of delivering the environmental benefits through efficient resource use and reduced environmental impacts (both on - and off-farm). It also help protect biodiversity and enhances the resilience against the shocks associated with accelerating climate change. Agroecology is the emerging scientific paradigm based on the recognition of the ecological principals when applied to agricultural systems. Some of these ecological principles have been practiced in traditional agriculture and some of them have come from new findings in ecological disciplines, such as biodiversity and soil food web.

Agroecology is time-tested and proven traditional way of farming that was evolved by farmers suiting to their diverse agro-climatic conditions. The experience of farmers around the world using agroecological methods has provided enough evidence of its economic, social and environmental benefits. Agroecological approaches have delivered increased food production and improved income for farmers, and enhanced food security and nutrition for the communities they feed. The input costs go down with time and farmers not only reduce the cost of cultivation but also increases the output.

Need for Agroecology

In agriculture there is no sudden shift from one system to the other. A system evolves with time. Farmers have been changing their agricultural systems from time to time and according to circumstances. There has been a good deal of diversity among agricultural systems. Different systems were evolved in different geographical settings by different cultures. Primitive agriculture and traditional agriculture encompassed enormous diversity of farming systems. Therefore the agriculture under these two broad categories is extremely resilient and sustainable, and ensures us a sustainable future.

Agroeology is inevitable to provide balanced environments, sustained yields, biologically mediated soil fertility and natural pest regulation through the design of diversified Agro-ecosystem and the use of low-input technologies.

Principles of agroecology revolve round an agroecosystem – community of plants, animals and microorganisms, the three functional biotic components, interacting with each other and among themselves and with the physico-chemical environment modified by farmers to produce foods, fodder, fiber, fuel, and other useful products. Agroecology provides us an opportunity to make a holistic understanding of the agroecosystems which we design and manage for food production. Since ages farmers had been protecting, conserving and augmenting natural resources, such as forests, grasslands, biodiversity in forests/agro-biodiversity, livestock, soil, water resources, and overall farming cultures through applying the principles of agroecology.

Agroecology helps us understand and maintain vital mineral cycles, biological processes, energy transformations, and socioeconomic relationships in an integrated manner. Agricultural strategies woven around the principles of agroecology look into local geographical and socioeconomic specificities, environmental and cultural specificities, and obeys people's traditions, such as food habits, festivities and their ethical and aesthetic values. Some of the strategies for diversification of agro-ecosystem are: Agroforestry Systems, Mixed farming/Polycultures, Cover Crop, Crop Rotations and Livestock.

Reduction of 7% in microbial biomass was observed due to only chemical fertilizer application, whereas 22% decline in microbial biomass was obtained when herbicides was applied along with chemical fertilizer. In different agro ecosystems under organic farming 4 and 63% enhancement in microbial activity was noticed under different crops of various.

So there are enough evidences to prove that agroecology is not only profitable, but is also environment friendly and sustainable. Studies carried out in diverse ecosystems through the globe show that by conserving biodiversity and adopting organic farming, small farmers can increase their production and incomes. Food sovereignty, food security and food safety requires an urgent change in policy from chemical intensification which is leading to debt and suicides to biodiversity intensification which is creating a living economy of food. In the past and even till date agroecology has spread and being applied by many farming communities around the world, primarily through a process of farmer- to-farmer knowledge sharing. Its main inputs and investments are information, best practices, knowledge of local conditions, and the natural resources of local eco-systems. It is not dependent on synthetic chemical fertilizers, pesticides or transgenic crops, which are not only expensive for small and marginal farmers but are also resource-depleting. Keeping in view the challenges like climate change mitigation and adaptation, as well as future food security; empowering the family farmers in agroecological approach could play a crucial role, as agroecology offers the prospect of sustainable food production to meet the needs of a still growing global population. Agroecolgical farming also helps reducing the GHG emissions from the agricultural sector and also building resilience to already unavoidable climate change, protecting biodiversity, and sustaining communities and rural livelihoods.

The challenge of feeding nine billion people by 2050 can be met without any problem through getting back to agroecological methods for increasing nutritional yields and reducing the environmental impact and addressing the food wastage/loss at different stages from farm to table *vis-a-vis* Market access for small hold family farms. Agroecological practices enhance on-farm fertility production. Malawian farmers call it a "fertilizer factory in the fields." These practices reduce farmers' reliance on external inputs and state subsidies. This, in turn, makes vulnerable smallholders less dependent on local retailers and moneylenders.

Ecological Strategy

Agroecologists study a variety of agroecosystems, and the field of agroecology is not associated with any one particular method of farming, whether it be organic, conventional, intensive or extensive. Furthermore, it is not defined by certain management practices, such as the use of natural enemies in place of insecticides, or polyculture in place of monoculture.

Additionally, agroecologists do not unanimously oppose technology or inputs in agriculture but instead assess how, when, and if technology can be used in conjunction with natural, social and human assets. Agroecology proposes a context- or site-specific manner of studying agroecosystems, and as such, it recognizes that there is no universal formula or recipe for the success and maximum well-being of an agroecosystem.

Instead, agroecologists may study questions related to the four system properties of agroecosystems: productivity, stability, sustainability and equitability. As opposed to disciplines that are concerned with only one or some of these properties, agroecologists see all four properties as interconnected and integral to the success of an agroecosystem. Recognizing that these properties are found on varying spatial scales, agroecologists do not limit themselves to the study of agroecosystems at any one scale: farm, community, or global.

Agroecologists study these four properties through an interdisciplinary lens, using natural sciences to understand elements of agroecosystems such as soil properties and plant-insect interactions, as well as using social sciences to understand the effects of farming practices on rural communities, economic constraints to developing new production methods, or cultural factors determining farming practices.

Ecosystems Agroecology

This approach is based in the hypotheses that the natural systems, with its stability and resilience, provide the best model to mimic if sustainability is the goal. Normally, ecosystems agroecology is not actively involved in social science; however, this school is essentially based on the belief that large-scale agriculture is inappropriate. The work of Steve Gliessman is prototypical of this approach.

Agronomic Ecology

The basic approach in this branch is derived mostly from agronomy, including the traditional agricultural production sciences. This approach also does not actively involve social sciences in the agroecological analysis, but uses social sciences to understand the processes by which agriculture became unsustainable. Chuck Francis, Richard Hardwood, Ricardo Salvador, and Matt Liebman are exemplars of this approach.

Ecological Political Economy

The driving force behind this form of agroecology is a political-economical critique of modern agriculture. The school believes that only radical changes in political economy and the moral economy of research will reduce the negative costs of modern agriculture.

The works of Miguel Altieri (ecosystem biologist), John Vandermeer (population ecologist), Richard Lewontin, and Richard Levins provide examples of this politically charged and socially-oriented version of agroecology.

Agro-population Ecology

This approach is derived from the science of ecology primarily based on population ecology, which over the past three decades has been displacing the ecosystems biology of Odum. Buttel explains the main difference between the two categories, saying that "the application of population ecology to agroecology involves the primacy not only of analyzing agroecosystems from the perspective of the population dynamics of their constituent species, and their relationships to climate and biogeochemistry, but also there is a major emphasis placed on the role of genetics." David Andow and Alison Power are cited as examples of professionals espousing this view.

Integrated Assessment of Multifunctional Agricultural Systems

This approach focuses on the multifunctionality of the landscape, instead of focusing solely on the agricultural enterprise. Agriculture and the food system are considered parts of an institutional complex that relates to and integrates with other social institutions. Scholars adopting this highly integrated approach, mostly Europeans, do not consider any one discipline the leader of agroecology.

Holon Agroecology

First introduced in 2007 by the soil scientist William T. Bland and the environmental sociologist Michael M. Bell of the University of Wisconsin–Madison, holon agroecology draws on Koestler's notion of a "holon" which is both part and whole and develops it with ideas of narrative, intentionality, and incompleteness or unfinalizability, within an ever-changing "ecology of contexts". In contrast to systems thinking, holon agroecology stresses seeing the agricultural endeavor as an unfinished accomplishment that is constantly adjusting itself to its many contexts and their conflicts and incommensurabilities. The farm holon represents a kind of "holding together" in order to persist through change, but a holding together that is never fully unified and worked out.

Applications of Agroecology

To emit a point of view about a particular way of farming, an agroecologist would first seek to understand the contexts in which the farm(s) is(are) involved. Each farm may be inserted in a unique combination of factors or contexts. Each farmer may have their own premises about the meanings of an agricultural endeavor, and these meanings might be different than those of agroecologists. Generally, farmers seek a configuration that is viable in multiple contexts, such as family, financial, technical, political, logistical, market, environmental, spiritual. Agroecologists want to understand the behavior of those who seek livelihoods from plant and animal increase, acknowledging the organization and planning that is required to run a farm.

Approaches to Agroecology

Agroecologists do not always agree about what agroecology is or should be in the long-term. Different definitions of the term agroecology can be distinguished largely by the specificity with which one defines the term "ecology", as well as the term's potential political connotations. Definitions of agroecology, therefore, may be first grouped according to the specific contexts within which they situate agriculture. Agroecology is defined by the OECD as "the study of the relation of agricultural crops and environment." This definition refers to the "-ecology" part of "agroecology" narrowly as the natural environment. Following this definition, an agroecologist would study agriculture's various relationships with soil health, water quality, air quality, meso- and micro-fauna, surrounding flora, environmental toxins, and other environmental contexts.

A more common definition of the word can be taken from Dalgaard., who refer to agroecology as the study of the interactions between plants, animals, humans and the environment within agricultural systems. Consequently, agroecology is inherently multidisciplinary, including factors from agronomy, ecology, sociology, economics and related disciplines. In this case, the "-ecology" portion of "agroecology is defined broadly to include social, cultural, and economic contexts as well." Francis also expand the definition in the same way, but put more emphasis on the notion of food systems.

Agroecology is also defined differently according to geographic location. In the global south, the term often carries overtly political connotations. Such political definitions of the term usually ascribe to it the goals of social and economic justice; special attention, in this case, is often paid to the traditional farming knowledge of indigenous populations. North American and European uses of the term sometimes avoid the inclusion of such overtly political goals. In these cases, agroecology is seen more strictly as a scientific discipline with less specific social goals.

Agro-population Ecology

This approach is derived from the science of ecology primarily based on population ecology, which over the past three decades has been displacing the ecosystems biology of Odum. Buttel explains the main difference between the two categories, saying that "the application of population ecology to agroecology involves the primacy not only of analyzing agroecosystems from the perspective of the population dynamics of their constituent species, and their relationships to climate and biogeochemistry, but also there is a major emphasis placed on the role of genetics."

Indigenous Agroecology

This concept was proposed by political ecologist Josep Garí to recognise and uphold the integrated agro-ecological practices of many indigenous peoples, who simultaneously and sustainably safeguard, manage and use ecosystems for agricultural, food, biodiversity and cultural purposes at the same time. Indigenous agroecologies are not systems and practices halted in time, but keep co-evolving with new knowledge and resources, such as that provided by development projects, research initiatives and agro-biodiversity exchanges. In fact, the first agro-ecologists were indigenous peoples that advocated development policies and programmes to support their systems, rather than replacing them.

Inclusive Agroecology

Rather than viewing agroecology as a subset of agriculture, Wojtkowski takes a more encompassing perspective. In this, natural ecology and agroecology are the major headings under ecology. Natural ecology is the study of organisms as they interact with and within natural environments. Correspondingly, agroecology is the basis for the land-use sciences. Here humans are the primary governing force for organisms within planned and managed, mostly terrestrial, environments.

As key headings, natural ecology and agroecology provide the theoretical base for their respective sciences. These theoretical bases overlap but differ in a major way. Economics has no role in the functioning of natural ecosystems whereas economics sets direction and purpose in agroecology.

Under agroecology are the three land-use sciences, agriculture, forestry, and agroforestry. Although these use their plant components in different ways, they share the same theoretical core.

Beyond this, the land-use sciences further subdivide. The subheadings include agronomy, organic farming, traditional agriculture, permaculture, and silviculture. Within this system of subdivisions, agroecology is philosophically neutral. The importance lies in providing a theoretical base hitherto lacking in the land-use sciences. This allows progress in biocomplex agroecosystems including the multi-species plantations of forestry and agroforestry.

Agroecological Restoration

Agroecological restoration is the practice of re-integrating natural systems into agriculture in order to maximize sustainability, ecosystem services, and biodiversity. This is one example of a way to apply the principles of agroecology to an agricultural system.

Farms cannot be restored to a purely natural state because of the negative economic impact on farmers, but returning processes, such as pest control to nature with the method of intercropping, allows a farm to be more ecologically sustainable and, at the same time, economically viable. Agroecological restoration works toward this balance of sustainability and economic viability because conventional farming is not sustainable over the long run without the integration of natural systems and because the use of land for agriculture has been a driving force in creating the present world biodiversity crisis. Its efforts are complementary to, rather than a substitute for, biological conservation.

Agriculture creates a conflict over the use of land between wildlife and humans. Though the domestication of crop plants occurred 10,000 years ago, a 500% increase in the amount of pasture and crop land over the last three hundred years has led to the rapid loss of natural habitats. In recent years, the world community acknowledged the value of biodiversity in treaties, such as the 1992 landmark Convention on Biological Diversity.

Reintegration

The reintegration of agricultural systems into more natural systems will result in decreased yield and produce a more complex system, but there will be considerable gains in biodiversity and ecosystem services.

Biodiversity

The Food and Agriculture Organization of the United Nations estimates that more than 40% of earth's land surface is currently used for agriculture. And because so much land has been converted to agriculture, habitat loss is recognized as the driving force in biodiversity loss (FAO). This biodiversity loss often occurred in two steps, as in the American Midwest, with the introduction of mixed farming carried out on small farms and then with the widespread use of mechanized farming and monoculture beginning after World War II. The decline in farmland biodiversity can now be traced to changes in farming practices and increased agricultural intensity.

Increasing Heterogeneity

Heterogeneity (here, the diversity or complexity of the landscape) has been shown to be associated with species diversity. For example, the abundance of butterflies has been found to increase with heterogeneity. One important part of maintaining heterogeneity in the spaces between different fields is made up of habitat that is not cropped, such as grass margins and strips, scrub along field boundaries, woodland, ponds, and fallow land. These seemingly unimportant pieces of land are crucial for the biodiversity of a farm. The presence of field margins benefits many different taxa: the plants attract herbivorous insects, will which attract certain species of birds and those birds will attract their natural predators. Also, the cover provided by the no cropped habitat allows the species that need a large range to move across the landscape.

Monoculture

In the absence of cover, species face a landscape in which their habitat is greatly fragmented. The isolation of a species to a small habitat that it can't safely wander from can create a genetic bottleneck, decreasing the resilience of the particular population, and be another factor leading to the decline of the total population of the species. Monoculture, the practice of producing a single crop over a wide area, causes fragmentation. In conventional farming, monoculture, such as with rotations of corn and soybean crops planted in alternating growing seasons, is used so that very high yields can be produced. After the mechanization of farming, monoculture became a standard practice in corn-beans rotation, and had broad implications for the long-term sustainability and biodiversity of farms. Whereas organic fertilizers, had kept the soil's nutrients fixed to the ecosystem, the introduction of monoculture removed the nutrients and farmers compensated for that loss by using inorganic fertilizers. It is estimated that humans have doubled the rate of nitrogen input into the nitrogen cycle, mostly since 1975. As a result, the biological processes that controlled the way crops used the nutrients changed and the leached nitrogen from farmland soils has become a source of pollution.

Organic Farming

Organic farming prohibits the use of synthetic chemicals in crop and livestock production. It often includes diverse crop rotations and provides non-cropped habitat for insects that provide ecosystem services, such as pest control and pollination. Organic farmers are encouraged to follow wildlife friendly practices, and as a result there is a great difference between the ecosystem services that similarly sized but distinctly managed organic farms provide.

Fertilizer Tree

Fertilizer trees are used in agroforestry to improve the condition of soils used for farming. They capture nitrogen from the air and put it in the soil through their roots and falling leaves. They can also bring nutrients from deep in the soil up to the surface for crops with roots that cannot reach that depth. Fertilizer trees are further useful for preventing fertilizer erosion, soil degradation and related desertification, and improving water usage for crops.

Sesbania, *Gliricidia*, *Tephrosia*, and *Faidherbia albida* are known as fertilizer trees. Tree lucerne or tagasaste (*Cytisus proliferus*) is able to fix more than 587 kg of nitrogen per hectare per year. It can increase maize yields from 1 ton per hectare per year to more than 10 tons ha/year in areas with more than 850mm of rain per year or a perched water table. Tree lucerne is also used to create and maintain terra preta.

Organic Farming

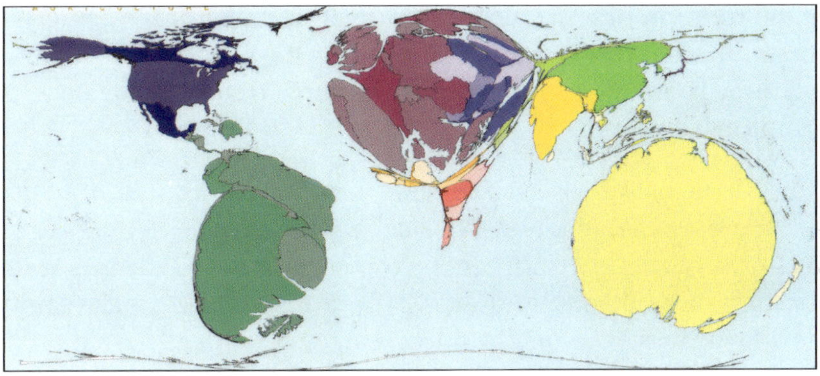

World map of organic agriculture (hectares).

Vegetables from ecological farming.

Organic farming is an alternative agricultural system which originated early in the 20th century in reaction to rapidly changing farming practices.

Agriculture was practiced for thousands of years without the use of artificial chemicals. Artificial fertilizers were first created during the mid-19th century. These early fertilizers were cheap, powerful, and easy to transport in bulk. Similar advances occurred in chemical pesticides in the 1940s, leading to the decade being referred to as the 'pesticide era'. These new agricultural techniques, while beneficial in the short term, had serious longer term side effects such as soil compaction, erosion, and declines in overall soil fertility, along with health concerns about toxic chemicals entering the food supply.

In the late 1800s and early 1900s, soil biology scientists began to seek ways to remedy these side effects while still maintaining higher production.

In 1921 the founder and pioneer of the organic movement Albert Howard and his wife Gabrielle Howard, accomplished botanists, founded an Institute of Plant Industry to improve traditional farming methods in India. Among other things, they brought improved implements and improved animal husbandry methods from their scientific training; then by incorporating aspects of the local traditional methods, developed protocols for the rotation of crops, erosion prevention techniques, and the systematic use of composts and manures. Stimulated by these experiences of traditional farming, when Albert Howard returned to Britain in the early 1930s he began to promulgate a system of organic agriculture.

In 1924 Rudolf Steiner gave a series of eight lectures on agriculture with a focus on influences of the moon, planets, non-physical beings and elemental forces. They were held in response to a request by adherent farmers who noticed degraded soil conditions and a deterioration in the health and quality of crops and livestock resulting from the use of chemical fertilizers. The one hundred eleven attendees, less than half of whom were farmers, came from six countries, primarily Germany and Poland. The lectures were published in November 1924; the first English translation appeared in 1928 as *The Agriculture Course*.

In July 1939, Ehrenfried Pfeiffer, the author of the standard work on biodynamic agriculture (*Bio-Dynamic Farming and Gardening*), came to the UK at the invitation of Walter James, 4th Baron Northbourne as a presenter at the Betteshanger Summer School and Conference on Biodynamic Farming at Northbourne's farm in Kent. One of the chief purposes of the conference was to bring together the proponents of various approaches to organic agriculture in order that they might cooperate within a larger movement. Howard attended the conference, where he met Pfeiffer. In the following year, Northbourne published his manifesto of organic farming, *Look to the Land*, in which he coined the term "organic farming." The Betteshanger conference has been described as the 'missing link' between biodynamic agriculture and other forms of organic farming.

In 1940, Howard published his *An Agricultural Testament*. In this book he adopted Northbourne's terminology of "organic farming." Howard's work spread widely, and he became known as the "father of organic farming" for his work in applying scientific knowledge and principles to various traditional and natural methods. In the United States J.I. Rodale, who was keenly interested both in Howard's ideas and in biodynamics, founded in the 1940s both a working organic farm for trials and experimentation, The Rodale Institute, and the Rodale Press to teach and advocate organic methods to the wider public. These became important influences on the spread of organic agriculture. Further work was done by Lady Eve Balfour (the Haughley Experiment) in the United Kingdom, and many others across the world.

Increasing environmental awareness in the general population in modern times has transformed the originally supply-driven organic movement to a demand-driven one. Premium prices and some government subsidies attracted farmers. In the developing world, many producers farm according to traditional methods that are comparable to organic farming, but not certified, and that may not include the latest scientific advancements in organic agriculture. In other cases, farmers in the developing world have converted to modern organic methods for economic reasons.

Terminology

The use of "organic" popularized by Howard and Rodale refers more narrowly to the use of organic matter derived from plant compost and animal manures to improve the humus content of soils, grounded in the work of early soil scientists who developed what was then called "humus farming." Since the early 1940s the two camps have tended to merge.

Biodynamic agriculturists, on the other hand, used the term "organic" to indicate that a farm should be viewed as a living organism, in the sense of the following quotation:

> "An organic farm, properly speaking, is not one that uses certain methods and substances and avoids others; it is a farm whose structure is formed in imitation of the structure of a natural system that has the integrity, the independence and the benign dependence of an organism."
>
> – *Wendell Berry*

They based their work on Steiner's spiritually-oriented alternative agriculture which includes various esoteric concepts.

Methods

Organic cultivation of mixed vegetables in Capay, California.

Organic agriculture is a production system that sustains the health of soils, ecosystems and people. It relies on ecological processes, biodiversity and cycles adapted to local conditions, rather than the use of inputs with adverse effects. Organic agriculture combines tradition, innovation and science to benefit the shared environment and promote fair relationships and a good quality of life for all involved.

Organic farming methods combine scientific knowledge of ecology and some modern technology with traditional farming practices based on naturally occurring biological processes. Organic farming methods are studied in the field of agroecology. While conventional agriculture uses synthetic pesticides and water-soluble synthetically purified fertilizers, organic farmers are restricted by regulations to using natural pesticides and fertilizers. An example of a natural pesticide is pyrethrin, which is found naturally in the Chrysanthemum flower. The principal methods of organic farming include crop rotation, green manures and compost, biological pest control, and mechanical cultivation. These measures use the natural environment to enhance agricultural productivity: legumes are planted to fix nitrogen into the soil, natural insect predators are encouraged, crops are rotated to confuse pests and renew soil, and natural materials such as potassium bicarbonate and mulches are used to control disease and weeds. Genetically modified seeds and animals are excluded.

While organic is fundamentally different from conventional because of the use of carbon based fertilizers compared with highly soluble synthetic based fertilizers and biological pest control instead of synthetic pesticides, organic farming and large-scale conventional farming are not entirely mutually exclusive. Many of the methods developed for organic agriculture have been borrowed by more conventional agriculture. For example, Integrated Pest Management is a multifaceted strategy that uses various organic methods of pest control whenever possible, but in conventional farming could include synthetic pesticides only as a last resort.

Crop Diversity

Organic farming encourages Crop diversity. The science of agroecology has revealed the benefits of polyculture (multiple crops in the same space), which is often employed in organic farming. Planting a variety of vegetable crops supports a wider range of beneficial insects, soil microorganisms, and other factors that add up to overall farm health. Crop diversity helps environments thrive and protects species from going extinct.

Soil Management

Organic farming relies heavily on the natural breakdown of organic matter, using techniques like green manure and composting, to replace nutrients taken from the soil by previous crops. This biological process, driven by microorganisms such as mycorrhiza, allows the natural production of nutrients in the soil throughout the growing season, and has been referred to as *feeding the soil to feed the plant.* Organic farming uses a variety of methods to improve soil fertility, including crop rotation, cover cropping, reduced tillage, and application of compost. By reducing tillage, soil is not inverted and exposed to air; less carbon is lost to the atmosphere resulting in more soil organic carbon. This has an added benefit of carbon sequestration, which can reduce green house gases and help reverse climate change .

Placard advocating organic food rather than global warming.

Plants need a large number of nutrients in various quantities to flourish. Supplying enough nitrogen and particularly synchronization so that plants get enough nitrogen at the when plants need it most, is a challenge for organic farmers. Crop rotation and green manure ("cover crops") help to provide nitrogen through legumes (more precisely, the family Fabaceae), which fix nitrogen from the atmosphere through symbiosis with rhizobial bacteria. Intercropping, which is sometimes used for insect and disease control, can also increase soil nutrients, but the competition between the legume and the crop can be problematic and wider spacing between crop rows is required. Crop residues can be ploughed back into the soil, and different plants leave different amounts of nitrogen, potentially aiding synchronization. Organic farmers also use animal manure, certain processed fertilizers such as seed meal and various mineral powders such as rock phosphate and green sand, a naturally occurring form of potash that provides potassium. Together these methods help to control erosion. In some cases pH may need to be amended. Natural pH amendments include lime and sulfur, but in the U.S. some compounds such as iron sulfate, aluminum sulfate, magnesium sulfate, and soluble boron products are allowed in organic farming.

Mixed farms with both livestock and crops can operate as ley farms, whereby the land gathers fertility through growing nitrogen-fixing forage grasses such as white clover or alfalfa and grows cash crops or cereals when fertility is established. Farms without livestock ("stockless") may find it more difficult to maintain soil fertility, and may rely more on external inputs such as imported manure as well as grain legumes and green manures, although grain legumes may fix limited nitrogen because they are harvested. Horticultural farms that grow fruits and vegetables in protected conditions often rely even more on external inputs.

Biological research into soil and soil organisms has proven beneficial to organic farming. Varieties of bacteria and fungi break down chemicals, plant matter and animal waste into productive soil nutrients. In turn, they produce benefits of healthier yields and more productive soil for future crops. Fields with less or no manure display significantly lower yields, due to decreased soil microbe community. Increased manure improves biological activity, providing a healthier, more arable soil system and higher yields.

Weed Management

Organic weed management promotes weed suppression, rather than weed elimination, by enhancing crop competition and phytotoxic effects on weeds. Organic farmers integrate cultural, biological, mechanical, physical and chemical tactics to manage weeds without synthetic herbicides.

Organic standards require rotation of annual crops, meaning that a single crop cannot be grown in the same location without a different, intervening crop. Organic crop rotations frequently include weed-suppressive cover crops and crops with dissimilar life cycles to discourage weeds associated with a particular crop.

Other cultural practices used to enhance crop competitiveness and reduce weed pressure include selection of competitive crop varieties, high-density planting, tight row spacing, and late planting into warm soil to encourage rapid crop germination.

Mechanical and physical weed control practices used on organic farms can be broadly grouped as:

- Tillage - Turning the soil between crops to incorporate crop residues and soil amendments; remove existing weed growth and prepare a seedbed for planting; turning soil after seeding to kill weeds, including cultivation of row crops;
- Mowing and cutting - Removing top growth of weeds;
- Flame weeding and thermal weeding - Using heat to kill weeds; and
- Mulching - Blocking weed emergence with organic materials, plastic films, or landscape fabric.

Some naturally sourced chemicals are allowed for herbicidal use. These include certain formulations of acetic acid (concentrated vinegar), corn gluten meal, and essential oils. A few selective bioherbicides based on fungal pathogens have also been developed. At this time, however, organic herbicides and bioherbicides play a minor role in the organic weed control toolbox.

Weeds can be controlled by grazing. For example, geese have been used successfully to weed a range of organic crops including cotton, strawberries, tobacco, and corn, reviving the practice of keeping cotton patch geese, common in the southern U.S. before the 1950s. Similarly, some rice farmers introduce ducks and fish to wet paddy fields to eat both weeds and insects.

Controlling other Organisms

Chloroxylon is used for Pest Management in Organic Rice Cultivation in Chhattisgarh, India.

Organisms aside from weeds that cause problems on organic farms include arthropods (e.g., insects, mites), nematodes, fungi and bacteria. Organic practices include, but are not limited to:

- Encouraging predatory beneficial insects to control pests by serving them nursery plants and/or an alternative habitat, usually in a form of a shelterbelt, hedgerow, or beetle bank;
- Encouraging beneficial microorganisms;
- Rotating crops to different locations from year to year to interrupt pest reproduction cycles;
- Planting companion crops and pest-repelling plants that discourage or divert pests;
- Using row covers to protect crops during pest migration periods;
- Using biologic pesticides and herbicides;
- Using stale seed beds to germinate and destroy weeds before planting;

- Using sanitation to remove pest habitat;
- Using insect traps to monitor and control insect populations; and
- Using physical barriers, such as row covers.

Examples of predatory beneficial insects include minute pirate bugs, big-eyed bugs, and to a lesser extent ladybugs (which tend to fly away), all of which eat a wide range of pests. Lacewings are also effective, but tend to fly away. Praying mantis tend to move more slowly and eat less heavily. Parasitoid wasps tend to be effective for their selected prey, but like all small insects can be less effective outdoors because the wind controls their movement. Predatory mites are effective for controlling other mites.

Naturally derived insecticides allowed for use on organic farms use include *Bacillus thuringiensis* (a bacterial toxin), pyrethrum (a chrysanthemum extract), spinosad (a bacterial metabolite), neem (a tree extract) and rotenone (a legume root extract). Fewer than 10% of organic farmers use these pesticides regularly; one survey found that only 5.3% of vegetable growers in California use rotenone while 1.7% use pyrethrum. These pesticides are not always more safe or environmentally friendly than synthetic pesticides and can cause harm. The main criterion for organic pesticides is that they are naturally derived, and some naturally derived substances have been controversial. Controversial natural pesticides include rotenone, copper, nicotine sulfate, and pyrethrums Rotenone and pyrethrum are particularly controversial because they work by attacking the nervous system, like most conventional insecticides. Rotenone is extremely toxic to fish and can induce symptoms resembling Parkinson's disease in mammals. Although pyrethrum (natural pyrethrins) is more effective against insects when used with piperonyl butoxide (which retards degradation of the pyrethrins), organic standards generally do not permit use of the latter substance.

Naturally derived fungicides allowed for use on organic farms include the bacteria *Bacillus subtilis* and *Bacillus pumilus*; and the fungus *Trichoderma harzianum*. These are mainly effective for diseases affecting roots. Compost tea contains a mix of beneficial microbes, which may attack or out-compete certain plant pathogens, but variability among formulations and preparation methods may contribute to inconsistent results or even dangerous growth of toxic microbes in compost teas.

Some naturally derived pesticides are not allowed for use on organic farms. These include nicotine sulfate, arsenic, and strychnine.

Synthetic pesticides allowed for use on organic farms include insecticidal soaps and horticultural oils for insect management; and Bordeaux mixture, copper hydroxide and sodium bicarbonate for managing fungi. Copper sulfate and Bordeaux mixture (copper sulfate plus lime), approved for organic use in various jurisdictions, can be more environmentally problematic than some synthetic fungicides dissallowed in organic farming. Similar concerns apply to copper hydroxide. Repeated application of copper sulfate or copper hydroxide as a fungicide may eventually result in copper accumulation to toxic levels in soil, and admonitions to avoid excessive accumulations of copper in soil appear in various organic standards and elsewhere. Environmental concerns for several kinds of biota arise at average rates of use of such substances for some crops.

Livestock

For livestock, like these healthy cows, vaccines play an important part in animal health since antibiotic therapy is prohibited in organic farming.

Raising livestock and poultry, for meat, dairy and eggs, is another traditional farming activity that complements growing. Organic farms attempt to provide animals with natural living conditions and feed. Organic certification verifies that livestock are raised according to the USDA organic regulations throughout their lives. These regulations include the requirement that all animal feed must be certified organic.

Organic livestock may be, and must be, treated with medicine when they are sick, but drugs cannot be used to promote growth, their feed must be organic, and they must be pastured.

Also, horses and cattle were once a basic farm feature that provided labor, for hauling and plowing, fertility, through recycling of manure, and fuel, in the form of food for farmers and other animals. While today, small growing operations often do not include livestock, domesticated animals are a desirable part of the organic farming equation, especially for true sustainability, the ability of a farm to function as a self-renewing unit.

Genetic Modification

A key characteristic of organic farming is the rejection of genetically engineered plants and animals. On 19 October 1998, participants at IFOAM's 12th Scientific Conference issued the Mar del Plata Declaration, where more than 600 delegates from over 60 countries voted unanimously to exclude the use of genetically modified organisms in food production and agriculture.

Organic farmer Raoul Adamchak and geneticist Pamela Ronald write that many agricultural applications of biotechnology are consistent with organic principles and have significantly advanced sustainable agriculture.

Although GMOs are excluded from organic farming, there is concern that the pollen from genetically modified crops is increasingly penetrating organic and heirloom seed stocks, making it difficult, if not impossible, to keep these genomes from entering the organic food supply. Differing regulations among countries limits the availability of GMOs to certain countries, as described in the topic on regulation of the release of genetic modified organisms.

Tools

Organic farmers use a number of traditional farm tools to do farming. Due to the goals of sustainability in organic farming, organic farmers try to minimize their reliance on fossil fuels. In the developing world on small organic farms tools are normally constrained to hand tools and diesel powered water pumps.

Standards

Standards regulate production methods and in some cases final output for organic agriculture. Standards may be voluntary or legislated. As early as the 1970s private associations certified organic producers. In the 1980s, governments began to produce organic production guidelines. In the 1990s, a trend toward legislated standards began, most notably with the 1991 EU-Eco-regulation developed for European Union, which set standards for 12 countries, and a 1993 UK program. The EU's program was followed by a Japanese program in 2001, and in 2002 the U.S. created the National Organic Program (NOP). As of 2007 over 60 countries regulate organic farming. In 2005 IFOAM created the Principles of Organic Agriculture, an international guideline for certification criteria. Typically the agencies accredit certification groups rather than individual farms.

Organic production materials used in and foods are tested independently by the Organic Materials Review Institute.

Composting

Using manure as a fertilizer risks contaminating food with animal gut bacteria, including pathogenic strains of E. coli that have caused fatal poisoning from eating organic food. To combat this risk, USDA organic standards require that manure must be sterilized through high temperature thermophilic composting. If raw animal manure is used, 120 days must pass before the crop is harvested if the final product comes into direct contact with the soil. For products that don't directly contact soil, 90 days must pass prior to harvest.

Disadvantages

A 2003 to 2005 investigation by the Cranfield University for the Department for Environment, Food and Rural Affairs in the UK found that it is difficult to compare the Global warming potential, acidification and eutrophication emissions but "Organic production often results in increased burdens, from factors such as N leaching and N_2O emissions", even though primary energy use was less for most organic products. N_2O is always the largest global warming potential contributor except in tomatoes. However, "organic tomatoes always incur more burdens (except pesticide use)". Some emissions were lower "per area", but organic farming always required 65 to 200% more field area than non-organic farming. The numbers were highest for bread wheat (200+ % more) and potatoes (160% more).

Environmental Impact and Emissions

Organic products are sometimes worse for the environment. Organic milk, cereals, and pork generated higher greenhouse gas emissions per product than conventional ones but organic beef and olives had lower emissions in most studies. Usually organic products required less energy, but more land. Per unit of product, organic produce generates higher nitrogen leaching, nitrous oxide emissions, ammonia emissions, eutrophication, and acidification potential than conventionally grown produce. Other differences were not significant.

Proponents of organic farming have claimed that organic agriculture emphasizes closed nutrient cycles, biodiversity, and effective soil management providing the capacity to mitigate and even reverse the effects of climate change and that organic agriculture can decrease fossil fuel emissions. "The carbon sequestration efficiency of organic systems in temperate climates is almost double (575–700 kg carbon per ha per year – 510–625 lb/ac/an) that of conventional treatment of soils, mainly owing to the use of grass clovers for feed and of cover crops in organic rotations."

Critics of organic farming methods believe that the increased land needed to farm organic food could potentially destroy the rainforests and wipe out many ecosystems.

Nutrient Leaching

According to a 2012 meta-analysis of 71 studies, nitrogen leaching, nitrous oxide emissions, ammonia emissions, eutrophication potential and acidification potential were higher for organic products, although in one study "nitrate leaching was 4.4–5.6 times higher in conventional plots than organic plots". Excess nutrients in lakes, rivers, and groundwater can cause algal blooms, eutrophication, and subsequent dead zones. In addition, nitrates are harmful to aquatic organisms by themselves.

Land Use

The Oxford meta-analysis of 71 studies found that organic farming requires 84% more land for an equivalent amount of harvest, mainly due to lack of nutrients but sometimes due to weeds, diseases or pests, lower yielding animals and land required for fertility building crops. While organic farming does not necessarily save land for wildlife habitats and forestry in all cases, the most modern breakthroughs in organic are addressing these issues with success.

Professor Wolfgang Branscheid says that organic animal production is not good for the environment, because organic chicken requires twice as much land as "conventional" chicken and organic pork a quarter more. According to a calculation by Hudson Institute, organic beef requires three times as much land. On the other hand, certain organic methods of animal husbandry have been shown to restore desertified, marginal, and/or otherwise unavailable land to agricultural productivity and wildlife. Or by getting both forage and cash crop production from the same fields simultaneously, reduce net land use.

In England organic farming yields 55% of normal yields. In other regions of the world, organic methods have started producing record yields.

Pesticides

A sign outside of an organic apple orchard in Pateros, Washington reminding orchardists not to spray pesticides on these trees.

In organic farming synthetic pesticides are generally prohibited. A chemical is said to be synthetic if it does not already exist in the natural world. But the organic label goes further and usually prohibit compounds that exist in nature if they are produced by chemical synthesis. So the prohibition is also about the method of production and not only the nature of the compound.

A non-exhaustive list of organic approved pesticides with their median lethal doses:

- Copper(II) sulfate is used as a fungicide and is also used in conventional agriculture (LD_{50} 300 mg/kg). Conventional agriculture has the option to use the less toxic Mancozeb (LD_{50} 4,500 to 11,200 mg/kg).
- Boric acid is used as stomach poison that target insects (LD_{50}: 2660 mg/kg).
- Pyrethrin comes from chemicals extracted from flowers of the genus Pyrethrum (LD_{50} of 370 mg/kg). Its potent toxicity is used to control insects.
- Lime sulfur (aka calcium polysulfide) and sulfur are considered to be allowed, synthetic materials (LD_{50}: 820 mg/kg).
- Rotenone is a powerful insecticide that was used to control insects (LD_{50}: 132 mg/kg). Despite the high toxicity of Rotenone to aquatic life and some links to Parkinson disease the compound is still allowed in organic farming as it is a naturally occurring compound.
- Bromomethane is a gas that is still used in the nurseries of Strawberry organic farming.
- Azadirachtin is a wide spectrum very potent insecticide. Almost non toxic to mammals (LD_{50} in rats is > 3,540 mg/kg) but affects beneficial insects.

Food Quality and Safety

While there may be some differences in the amounts of nutrients and anti-nutrients when organically produced food and conventionally produced food are compared, the variable nature of food production and handling makes it difficult to generalize results, and there is insufficient evidence to make claims that organic food is safer or healthier than conventional food. Claims that organic food tastes better are not supported by evidence.

Soil Conservation

Supporters claim that organically managed soil has a higher quality and higher water retention. This may help increase yields for organic farms in drought years. Organic farming can build up soil organic matter better than conventional no-till farming, which suggests long-term yield benefits from organic farming. An 18-year study of organic methods on nutrient-depleted soil concluded that conventional methods were superior for soil fertility and yield for nutrient-depleted soils in cold-temperate climates, arguing that much of the benefit from organic farming derives from imported materials that could not be regarded as self-sustaining.

In *Dirt: The Erosion of Civilizations*, geomorphologist David Montgomery outlines a coming crisis from soil erosion. Agriculture relies on roughly one meter of topsoil, and that is being depleted ten times faster than it is being replaced. No-till farming, which some claim depends upon pesticides, is one way to minimize erosion. However, a 2007 study by the USDA's Agricultural Research Service has found that manure applications in tilled organic farming are better at building up the soil than no-till.

Biodiversity

The conservation of natural resources and biodiversity is a core principle of organic production. Three broad management practices (prohibition/reduced use of chemical pesticides and inorganic fertilizers; sympathetic management of non-cropped habitats; and preservation of mixed farming) that are largely intrinsic (but not exclusive) to organic farming are particularly beneficial for farmland wildlife. Using practices that attract or introduce beneficial insects, provide habitat for birds and mammals, and provide conditions that increase soil biotic diversity serve to supply vital ecological services to organic production systems. Advantages to certified organic operations that implement these types of production practices include: 1) decreased dependence on outside fertility inputs; 2) reduced pest management costs; 3) more reliable sources of clean water; and 4) better pollination.

Nearly all non-crop, naturally occurring species observed in comparative farm land practice studies show a preference for organic farming both by abundance and diversity. An average of 30% more species inhabit organic farms. Birds, butterflies, soil microbes, beetles, earthworms, spiders, vegetation, and mammals are particularly affected. Lack of herbicides and pesticides improve biodiversity fitness and population density. Many weed species attract beneficial insects that improve soil qualities and forage on weed pests. Soil-bound organisms often benefit because of increased bacteria populations due to natural fertilizer such as manure, while experiencing reduced intake of herbicides and pesticides. Increased biodiversity, especially from beneficial soil microbes and mycorrhizae have been proposed as an explanation for the high yields experienced by some organic plots, especially in light of the differences seen in a 21-year comparison of organic and control fields.

Biodiversity from organic farming provides capital to humans. Species found in organic farms enhance sustainability by reducing human input (e.g., fertilizers, pesticides).

The USDA's Agricultural Marketing Service (AMS) published a *Federal Register* notice on 15 January 2016, announcing the National Organic Program (NOP) final guidance on Natural Resources and Biodiversity Conservation for Certified Organic Operations. Given the broad scope of natural resources which includes soil, water, wetland, woodland and wildlife, the guidance provides examples of practices that support the underlying conservation principles and demonstrate compliance with USDA organic regulations. The final guidance provides organic certifiers and farms with examples of production practices that support conservation principles and comply with the USDA organic regulations, which require operations to maintain or improve natural resources. The final guidance also clarifies the role of certified operations (to submit an OSP to a certifier), certifiers (ensure that the OSP describes or lists practices that explain the operator's monitoring plan and practices to support natural resources and biodiversity conservation), and inspectors (onsite inspection) in the implementation and verification of these production practices.

A wide range of organisms benefit from organic farming, but it is unclear whether organic methods confer greater benefits than conventional integrated agri-environmental programs. Organic farming is often presented as a more biodiversity-friendly practice, but the generality of the beneficial effects of organic farming is debated as the effects appear often species- and context-dependent, and current research has highlighted the need to quantify the relative effects of local- and landscape-scale management on farmland biodiversity. There are four key issues when comparing the impacts on biodiversity of organic and conventional farming: (1) It remains unclear whether a holistic whole-farm approach (i.e. organic) provides greater benefits to biodiversity than carefully targeted prescriptions applied to relatively small areas of cropped and/or non-cropped habitats within conventional agriculture; (2) Many comparative studies encounter methodological problems, limiting their ability to draw quantitative conclusions; (3) Our knowledge of the impacts of organic farming in pastoral and upland agriculture is limited; (4) There remains a pressing need for longitudinal, system-level studies in order to address these issues and to fill in the gaps in our knowledge of the impacts of organic farming, before a full appraisal of its potential role in biodiversity conservation in agroecosystems can be made.

Opposition to Labor Standards

Organic agriculture is often considered to be more socially just and economically sustainable for farmworkers than conventional agriculture. However, there is little social science research or consensus as to whether or not organic agriculture provides better working conditions than conventional agriculture. As many consumers equate organic and sustainable agriculture with small-scale, family-owned organizations it is widely interpreted that buying organic supports better conditions for farmworkers than buying with conventional producers. Organic agriculture is generally more labor-intensive due to its dependence on manual practices for fertilization and pest removal and relies heavily upon hired, non-family farmworkers rather than family members. Although illnesses from synthetic inputs pose less of a risk, hired workers still fall victim to debilitating musculoskeletal disorders associated with agricultural work. The USDA certification requirements outline growing practices and ecological standards but do nothing to codify labor practices. Independent certification initiatives such as the Agricultural Justice Project, Domestic Fair Trade Working Group, and the Food Alliance have attempted to implement farmworker interests but because these initiatives require voluntary participation of organic farms, their standards cannot be widely enforced. Despite the benefit to farmworkers of implementing labor standards, there is little support among the organic community for these social requirements. Many actors of the organic industry believe that enforcing labor standards would be unnecessary, unacceptable, or unviable due to the constraints of the market.

Bioeffector

A Bioeffector is a viable microorganism or active natural compound which directly or indirectly affects plant performance (Biofertilizer), and thus has the potential to reduce fertilizer and pesticide use in crop production.

Types

Bioeffectors have a direct or indirect effect on plant performance by influencing the functional implementation or activation of biological mechanisms, particularly those interfering with soil-plant-microbe interactions. In contrast to conventional fertilizers and pesticides, the effectiveness of bioeffectors is not based on a substantial direct input of mineral plant nutrients, either in inorganic or organic forms.

- Products in use are:
  - Microbial residues,
  - Composting and fermentation products,
  - Plant and algae extracts.
- Bioeffector-preparations (*bio-agents*) as ready-formulated products are applied:
  - With the purpose of stimulating plant growth (bio-stimulants),
  - To improve plant nutrient acquisition (bio-fertilizers),
  - To protect plants from pathogens and pests (bio-control agents),
  - Or generally to advance cropping efficiency; they can contain one or more bio-effectors along with other substances.
- Well established bioeffectors with documented positive results in the field level are:
  - Rhizobia strains for soil or seed inoculation as a prerequisite for symbiotic N2-fixation when establishing new legume species or varieties.

  - Positive effects of mycorrhiza inoculation for soils with a (temporarily) low potential for natural root mycorrhization.
  - Sufficient mycorrhization enhances nutrient (P) and water uptake and increases resistance to pathogenic fungi.
- Further mechanisms for the positive impact of bioeffectors on plant growth have postulated, promising a high potential for resource preservation due to reduction of fertiliser and pesticide use:
  - Active nutrient mobilisation by exudation of acids and carboxylates (e.g. P-mobilisation),
  - Exudation of micro-nutrient mobilising siderophores/chelates (e.g. Fe^{3+}),
  - Reduction of trace elements from less soluble oxidised to highly soluble reduced forms (e.g. Fe^{3+} to Fe^{2+}, Mn^{4+} to Mn^{2+}),
  - Associative/non-symbiotic N_2-fixation, protective antagonism to plant pathogens,
  - Enhancement of mycorrhizal infection and growth, and stimulating hormonal effects.

Good Agricultural Practice

Good agricultural practice (GAP) are specific methods which, when applied to agriculture, create food for consumers or further processing that is safe and wholesome. While there are numerous competing definitions of what methods constitute good agricultural practice there are several broadly accepted schemes that producers can adhere to.

Food and Agricultural Organization of the United Nations GAP

The Food and Agricultural Organization of the United Nations (FAO) uses good agricultural practice as a collection of principles to apply for on-farm production and post-production processes, resulting in safe and healthy food and non-food agricultural products, while taking into account economical, social and environmental sustainability.

GAPs may be applied to a wide range of farming systems and at different scales. They are applied through sustainable agricultural methods, including economically and efficiently produce being sufficient (food security), safe (food safety), and making sure that the food is nutritious (food quality).

GAPs require maintaining a common database on integrated production techniques for each of the major agro-ecological area. They collect, analyze and disseminate information of good practices in relevant geographical contexts.

Smallholder Productivity

Demand for agricultural crops is expected to double as the world's population reaches 9.1 billion by 2050. Increasing the quantity and quality of food in response to growing demand will require increased agricultural production. Good agricultural practices, often in combination with effective input use, are one of the best ways to increase smallholder productivity. Many agribusinesses are building sustainable supply chains to increase production and improve quality.

Soil

- Reducing erosion by wind and water through hedging and ditching;
- Application of fertilizers at appropriate moments and in adequate doses (i.e., when the plant needs the fertilizer), to avoid run-off;
- Maintaining or restoring soil organic content, by manure application, use of grazing, crop rotation;
- Reduce soil compaction issues (by avoiding using heavy mechanical devices);
- Maintain soil structure, by limiting heavy tillage practices;
- *In situ* green manuring by growing pulse crops like cowpea, horse gram, sunn hemp etc.

Water

- Practice scheduled irrigation, with monitoring of plant needs, and soil water reserve status to avoid water loss by drainage;
- Prevent soil salinization by limiting water input to needs, and recycling water whenever possible;
- Avoid crops with high water requirements in a low availability region;
- Avoid drainage and fertilizer run-off;
- Maintain permanent soil covering, in particular in winter to avoid nitrogen run-off;
- Manage carefully water table, by limiting heavy output of water;
- Restore or maintain wetlands;

- Provide good water points for livestock;
- Harvest water *in situ* by digging catch pits, crescent bunds across slope.

Animal Production, Health and Welfare

- Respect of animal well-being (freedom from hunger and thirst; freedom from discomfort; freedom from pain, injury or disease; freedom to express normal behavior; and freedom from fear and distress);
- Avoid nontherapeutic mutilations, surgical or invasive procedures, such as tail docking and debeaking;
- Avoid negative impacts on landscape, environment and life: contamination of land for grazing, food, water and air;
- Check stocks and flows, maintain structure of systems;
- Prevent chemical and medical residues from entering the food chain;
- Minimize non-therapeutic use of antibiotics or hormones;
- Avoid feeding animals with animal wastes or animal matter (reducing the risk of alien viral or transgenic genes, or prions such as mad cow disease);
- Minimize transport of live animals (by foot, rail or road) (reducing the risk of epidemics, e.g., foot and mouth disease);
- Prevent waste run-off (e.g. nitrate contamination of water tables from pigs), nutrient loss and greenhouse gas emissions (methane from cows);
- Prefer safety measures standards in manipulation of equipment;
- Apply traceability processes on the whole production chain (breeding, feed, medical treatment) for consumer security and feedback possibility in case of a food crisis (e.g., dioxin).

Healthcare and Public Health

Quality assurance of the horticultural or agricultural production of medicinal plant.

The farm is the first step in the farm-to-consumer farm marketing system. Conventional or organic, the safety of the food system begins with your farm. Understand your responsibility to grow food using GAP and get a basic understanding of what these practices are by viewing the video and checking the list.

The produce industry's best practices include:

- Making sure the farm has well-maintained toilets and hand-washing facilities with potable water & hand soap for hand washing.

- Making sure employees are washing their hands before harvesting and handling produce.
- Using a proactive pest management strategy for rodents, birds, deer, pigs, slugs and snails.
- Keeping animals and their fresh manures away from active fields and orchards.
- Using the right crop protection chemicals, fertilizers and composts according to their labeled (legal) directions, and recording every use. And, educating growers on how to follow the US EPA Worker Protection Standard rules.
- Making sure that produce, and harvest baskets with holes, do not come in contact with the soil or unclean surfaces.
- Using appropriate quality water for irrigation and crop rinse as indicated by a water test from an approved laboratory.
- Making sure the packing shed, food contact packing surfaces, and refrigerators are well maintained and not a potential source of contamination.
- Labeling each sell unit with farmer contact information, "Grown in Hawaii," and the appropriate field and harvest information to allow trace-back to a specific field within 2 hours.

The guidelines are applicable to the production, harvesting, storage, etc. of local vegetables (including leafy, gourd and fruit vegetables, roots and rhizomes, legumes). Farmers can also refer to the "Good Agricultural Practices – Crop Production" Code of Practices and adopt appropriate measures to produce safe and premium vegetables.

Seed and Seedling

- Look for varieties which are pest and disease resistant, heat resistant and GM-free.
- Keep seeds in sealed containers and store in a cool and dry place.

Soil Management

- Frequently apply compost or other organic materials (including crop residues) and plant cover crops and/or green manure crops and minimize mechanical movement in fields.
- If necessary, acid soil can be corrected by liming, whereas alkaline soil can be corrected by sulphur.
- If necessary, saline soil can be rectified by leaching soil, scraping off surface soil and reducing use of agro-chemicals.

Use of Fertilizers

- Apply the required level of fertilizers at the right time.
- Use organic fertilizers and aged/well composted manure.
- Keep fertilizers in a dry, clean and sheltered place.
- Do not use empty/used fertilizer bags for harvested vegetables.

Irrigation

- Irrigate fields early in the morning, late in the evening or at night.
- Avoid uneven application of water.
- Adopt micro-irrigation methods such as drip or sprinkler.

Pest and Disease Management

- Always use preventive methods. Examples are using disease-free seeds/seedlings, adopting crop rotation and intercropping, planting green manure crops, hedges and crops with pest deterring value, and instant removal of infected/diseased materials.
- Adopt physical control measures. Examples include simple hand-picking, erecting traps, vacuuming, tillage, flaming, mulching, soil solarization and using net houses and greenhouses.
- If really necessary, use bio-pesticides/synthetic pesticides.

Use of Pesticides

- Only purchase and use registered pesticides.
- Do not apply pesticides during strong winds and heavy rain.
- Strictly adhere to the withholding period (i.e. the lag between pesticide application and harvesting) on the pesticide label.
- Hold pesticides in original containers and keep them tightly closed in a cool, well-ventilated location.
- Do not recycle or re-use pesticide containers for other usage.

Harvesting and Post-harvesting

- Harvest at the right stage of maturity.
- Harvest during the coolest part of the day - either early morning or late afternoon.

- If necessary, harvested produce should be washed with clean water and excessive water should be removed before packing or storing.
- Wash hands with soap before and after handling produce. Do not smoke, drink or eat during harvesting.
- Always keep containers, tools, equipment, packing and storage areas clean and tidy.

Regenerative Agriculture

Biodiversity.

Hoverfly at work.

Regenerative agriculture is a conservation and rehabilitation approach to food and farming systems. It focuses on topsoil regeneration, increasing biodiversity, improving the water cycle, enhancing ecosystem services, supporting biosequestration, increasing resilience to climate change, and strengthening the health and vitality of farm soil. Practices include, recycling as much farm waste as possible, and adding composted material from sources outside the farm.

Regenerative agriculture on small farms and gardens is often based on ideologies like permaculture, agroecology, agroforestry, restoration ecology, keyline design and holistic management. Large farms tend to be less ideology driven, and often use "no-till" and/or "reduced till" practices.

On a regenerative farm, yield should increase over time. As the topsoil deepens, production may increase and less external compost inputs are required. Actual output is dependent on the nutritional value of the composting materials, and the structure and content of the soil.

Principles and Practices

Regenerative agriculture is guided by a set of principles and practices.

Principles

Principles include:

- Increase soil fertility to nutrient saturation points.
- Work with whole systems, not isolated parts.
- Progressively improve whole agro-ecosystems (soil, water and biodiversity).
- Connect the farm to its larger agro-ecosystem and bio-region.
- Create designs and make holistic decisions that express the value and relationship of farm contributors.
- Express the significance of each person, farm and system.
- Make holistic decisions aimed at specific system changes.
- Ensure and develop equitable and reciprocal relationships among all stakeholders.
- Design for non-linear, multi-capital reciprocity.
- Continually grow and evolve individuals, farms and communities to express their potential.
- Continuously evolve agro-ecological processes and cultures.
- Agriculture influences the world.

Practices

Practices include but are not limited to:

- Permaculture Design;
- Agroforestry;
- Soil Food Web;
- Properly Managed Livestock, Well-managed grazing, Animal Integration, and Holistically Managed Grazing;
- STUN (Sheer, Total and Utter Neglect) Breeding;
- Keyline Subsoiling;
- Conservation farming, No-Till Farming, minimum tillage, and Pasture Cropping;
- Cover crops & multi-species cover crops;
- Organic Annual Cropping and Crop rotations;
- Compost, Compost Tea, animal manures and Thermal Compost;
- Natural sequence farming;
- Grassfed livestock;
- Polyculture and full-time succession planting of multiple and inter-crop plantings;
- Borders planted for pollinator habitat and other beneficial insects;
- Biochar/Terra Preta;
- Ecological Aquaculture;
- Perennial Crops;
- Silvopasture.

Carbon Farming

Carbon farming is a name for a variety of agricultural methods aimed at sequestering atmospheric carbon into the soil. Increasing the carbon content of soil can aid plant growth, increase soil organic matter (improving agricultural yield), improve soil water retention capacity, and reduce fertilizer use (and the accompanying emissions of greenhouse gas nitrous oxide (N_2O). As of 2016, variants of carbon farming affected hundreds of millions of hectares globally, of the nearly 5,000,000,000 hectares (1.2×10^{10} acres) of world farmland. Soils can contain up to five per cent carbon by weight, including decomposing plant and animal matter and biochar.

Potential sequestration alternatives to carbon farming include scrubbing the air with machines; fertilizing the oceans to prompt algal blooms that after death carry carbon to the sea bottom; storing the carbon dioxide emitted by electricity generation; and crushing and spreading types of rock such as basalt that absorb atmospheric carbon. Land management techniques that can be combined with farming include planting/restoring forests, burying biochar produced by anaerobically converted biomass and restoring wetlands. (Coal beds are the remains of marshes and peatlands.)

Soil Carbon

Traditionally, soil carbon was thought to accumulate when decaying organic matter was physically mixed with soil. More recently, the role of living plants has been emphasized. Small roots die and decay while the plant is alive, depositing carbon below the surface. Further, as plants grow, their roots inject some carbon into the soil, feeding mycorrhiza. An estimated 12,000 miles of their hyphae live in every square meter of healthy soil.

Techniques

At least thirty-two Natural Resource Conservation Service (NRCS) practices improve soil health and sequester carbon, along with important co-benefits: increased water retention, hydrological function, biodiversity and resilience. Approved practices may make farmers eligible for federal funds. Not all carbon farming techniques (e.g., composting) have been recommended. Carbon farming may consider related issues such as groundwater and surface water degradation.

Tilling

Carbon farming minimizes disruption to soils over the planting/growing/harvest cycle. Tillage is avoided using seed drills or similar techniques. Livestock can trample and/or eat the remains of a harvested field.

Livestock Grazing

Grazing livestock sequesters carbon when the animal eats the grass, causing its roots to release carbon into the soil. However, these animals also produce significant methane, potentially offsetting the benefits of sequestration. Livestock is regularly rotated through multiple paddocks (as often as daily) allowing the paddocks to rest/recover between grazing periods. This pattern produces stable grasslands with significant fodder. Annual grasses have shallower roots and die once they're grazed. Rotational grazing led to the replace of the annuals perennials with deeper roots, which can recover after limited grazing. By contrast, allowing animals to range over a large area for an extended period can destroy the grassland.

Silvopasture

Silvopasture involves grazing livestock under tree cover, with trees separated enough to allow adequate sunlight to reach the ground. For example, a farm in Mexico planted native trees on 22 hectares (54 acres) grazing cattle on the paddock. This evolved into a successful organic dairy. The operation became a subsistence farm, earning income from consulting/training others rather than from crop production.

Organic Mulch

Mulching covers the soil around plants with a mulch of wood chips or straw. Alternatively, crop residue can be left in place to enter the soil as it decomposes.

Compost

Compost sequesters carbon in a stable (not easily accessed) form. Carbon farmers dust it over the soil surface without tilling. A 2013 study found that a single compost application significantly and durably increased grassland carbon storage by 25-70%. The continuation sequestration likely came from increased water-holding and "fertilization" by compost decomposition. Both factors support increased productivity. Both tested sites showed large increases in grassland productivity: a forage increase of 78% in the drier valley site, while the wetter coastal site averaged an increase of 42%. CH_4 and N_2O and emissions did not increase significantly. Methane fluxes were negligible. Soil N_2O emissions from temperate grasslands amended with chemical fertilizers and manures were orders of magnitude higher. Another study found that grasslands treated with a single treatment .5" of commercial compost began absorbing carbon at an annual rate of nearly 1.5 tons/acre and continued to do so in subsequent years.

Cover Crops

With row crops such as corn and wheat, fast-growing ground cover is grown between the stalks (e.g., clover or vetch). They protect the soil from carbon loss through the winter and may be planted together with cash crops to compensate for carbon lost when those crops are harvested. Forage crops such as grasses, clovers and alfalfa develop extensive root systems that can become soil organic matter. Crops with poor root systems (corn, soybeans) do not increase organic matter in the soil.

Bamboo

Bamboo stores large amounts of carbon in the soil and in its superstructure.

Hybrids

Perennial crops offer the highest potential to sequester carbon, when grown in multilayered systems. These systems can be challenging and require farmers to cultivate potentially unfamiliar crops. One system uses perennial staple crops that grow on trees that are analogs to maize and beans, or vines, palms and herbaceous perennials.

Conventional Agriculture

Plowing splits soil aggregates and allows microorganisms to consume their organic compounds. The increased microbial activity releases nutrients, initially boosting yield. Thereafter the loss of structure reduces soil's ability to hold water and resist erosion, thereby reducing yield.

Criticisms

Critics say that the related regenerative agriculture cannot be adopted enough to matter or that it could lower commodity prices. The impact of increased soil carbon on yield has yet to be resolved. Another criticism says that no-till practices may increase herbicide use, diminishing or eliminating the benefits. Composting is not an NRCS-approved technique and its impacts on native species and the greenhouse emissions during production have not been fully resolved. Further, commercial compost supplies are too limited to cover large amounts of land.

References

- 296-what-is-agroecology, what-is-ecology: environment-ecology.com, Retrieved 31 March, 2019
- Funes, fernando; garcía, luis; bourque, martin; pérez, nilda; rosset, peter (january 2002). Sustainable agriculture and resistance: transforming food production in cuba. Oakland, ca: food first books. Isbn 978-0-935028-87-4
- What-is-agroecology: navdanya.org, Retrieved 14 July, 2019
- Garí, josep a. (2004). Plant diversity, sustainable rural livelihoods and the hiv/aids crisis. Bangkok: undp & fao, 2004. Published in english and chinese. Isbn 974-92021-4-7
- Hole, d.g.; perkins, a.j.; wilson, j.d.; alexander, i.h.; grice, p.v.; evans, a.d. (2005-03-01). "does organic farming benefit biodiversity?". Biological conservation. 122 (1): 113–130. Doi:10.1016/j.biocon.2004.07.018. Issn 0006-3207
- Langford, kate (october 14, 2011). "new study finds 400,000 farmers in southern africa using 'fertilizer trees' to dramatically improve food security". World agroforestry centre. Archived from the original on march 9, 2014. Retrieved august 29, 2012
- Step-1-basic-good-agricultural-practices, farmfoodsafety: manoa.hawaii.edu, Retrieved 17 May, 2019
- Falk, ben (2013). The resilient farm and homestead: an innovative permaculture and whole systems design approach. Chelsea green publishing. P. 280. Isbn 978-1-60358-444-9

Agricultural Biotechnology 3

- Crop Modification Techniques
- Improved Nutritional Content
- Genes and Traits of Interest for Crops
- Genetic Engineering in Agriculture

The field of agricultural science which is concerned with the utilization of scientific tools and techniques such as genetic engineering and molecular markers is known as agricultural biotechnology. Some of the crop modification techniques which are used within this field are traditional breeding, mutation breeding, protoplast fusion, transgenics, etc. This chapter closely examines these techniques of agricultural biotechnology.

Agricultural biotechnology is a collection of scientific techniques used to improve plants, animals and microorganisms. Based on an understanding of DNA, scientists have developed solutions to increase agricultural productivity. Starting from the ability to identify genes that may confer advantages on certain crops, and the ability to work with such characteristics very precisely, biotechnology enhances breeders' ability to make improvements in crops and livestock. Biotechnology enables improvements that are not possible with traditional crossing of related species alone.

Early farmers selected the best looking plants and seeds and saved them to plant for the next year. The selection for features such as faster growth, higher yields, pest and disease resistance, larger seeds, or sweeter fruits has dramatically changed domesticated plant species compared to their wild relatives. Plant breeding came into being when man learned that crop plants could be artificially mated or cross-pollinated to be able to improve the characters of the plant. Desirable characteristics from different parent plants could be combined in the offspring. When the science of plant breeding was further developed in the 20th century, plant breeders understood better how to select superior plants and breed them to create new and improved varieties of different crops. This has dramatically increased the productivity and quality of the plants we grow for food, feed and fiber.

Conventional plant breeding has been the method used to develop new varieties of crops for hundreds of years. However, conventional plant breeding can no longer sustain the global demand with the increasing population, decline in agricultural resources such as land and water, and the apparent plateauing of the yield curve of the staple crops. Thus, new crop improvement technologies should be developed and utilized.

Conventional breeding entails sexual hybridization followed by careful selection.

Mutation Breeding

The art of recognizing desirable traits and incorporating them into future generations is very important in plant breeding. Breeders inspect their fields and travel long distances in search of individual plants that exhibit desirable traits.

A few of these traits occasionally arise spontaneously through a process called mutation, but the natural rate of mutation is very slow and unreliable to produce plants that breeders would like to see.

In the late 1920s, researchers discovered that they could greatly increase the number of these variations or mutations by exposing plants to X-rays and mutation-inducing chemicals. "Mutation breeding" accelerated after World War II, when the techniques of the nuclear age became widely available. Plants were exposed to gamma rays, protons, neutrons, alpha particles, and beta particles to see if these would induce useful mutations. Chemicals such as sodium azide and ethyl methanesulphonate, were also used to cause mutations. Mutation breeding efforts continue around the world today. In the 73 years of mutation breeding (1939-2013), a total of 3,218 varieties obtained through mutation breeding have been registered in the IAEA database. Staple crops such as rice has registered 824 varieties, barley (312), wheat (274), maize (96), common bean (57), tomato (20), potato (16), sugarcane (13), soybean (2), as well as other important crops that were improved to possess agronomically-desirable charateristics.

Pure Line and Hybrid Seed Technology

The end result of plant breeding is either an open-pollinated (OP for corn) or inbred (for rice) varieties or an F1 (first filial generation) hybrid variety. OP and inbred varieties, when maintained and properly selected and produced, retain the same characteristics when multiplied.

Hybrid seeds are an improvement over OP and inbred seeds in terms of yield, resistance to pests and diseases, and time to maturity.

Hybrid seeds are developed by the hybridization or crossing of diversely-related parent lines. Pure lines are offsprings of several cycles of repeated self-pollination that "breed true" or produce sexual offspring that closely resemble their parents.

Pure line development involves firstly, the selection of lines in the existing germplasm which express the desired characteristics such as resistance to pest and diseases, early maturity, yield, and others. These traits may not be present in only one line, thus selected lines are bred together by hand. In self-pollinated plants, flowers are emasculated by removing the anthers or the male part of the flower by hand, and are pollinated by pollen from another line. The female parent is usually the line that possesses the desired agronomic trait while the male parent is the donor of the new trait. F1 (first filial generation) offsprings are planted and selfed, as well as the F2 generation. Breeders then select in the F3 and F4 generation the lines which exhibit their desired agronomic characteristics and the added trait. Testing for resistances to pests and abiotic stresses are conducted also at this time. Lines with desired traits and are rated intermediate to resistant/tolerant to the pests and abiotic stresses are selected and selfed in two to three more generations. Lines which do not lose the new traits and are stable are termed pure lines.

In hybrid seed technology, two pure lines with complementing traits and are derived from diversely related parents are bred together by hand. F1 hybrids are tested for hybrid vigor in all agronomic and yield parameters and compared to both parents. The resulting offsprings will usually perform more vigorously than either parents.

Since the technology has been developed, it has brought tremendous impact in major crops including rice, corn, wheat, cotton, and other crops including many vegetables. In the USA, corn yield from 1866 to 1936 was only 26 bu/acre. Adoption of hybrid corn has increased corn yield by 0.8 bu/ac/yr from 1947-1955. With improved genetics, availability of N fertilizer, chemical pesticide and mechanization, corn grain yield has constantly increased by 1.9 bushels/acre/year to become 115 bushels in 1990's to an expected increase of 159 bu/acre in 2012. However, with the Great Drought in the US in 2012, grain yield was only 123.4 bu/acre. In 2013, an increase of 50 bu/acre of corn yield was obtained.

Hybrid rice technology helped China to increase production from 140 million tons in 1978 to 188 million tons in 1990. Since then, hybrid rice has helped increase rice production which yields 1.35 to 2 tons/hectare more than the ordinary rice, and hence an average yield of 7.2 to 7.5 tons/hectare. Hybrid rice production area is expected to increase by more than 6 million hectares in 2012. In September 2012, Yuan Long-pin, the farther of hybrid rice has completed the development of super rice DH2525 that sets a new record of hybrid rice yield at 926.6 kg/mu.

During the 6th Hybrid Rice Symposium in India in September 2012, Indian government and scientists realized the country's need to increase hectarage of hybrid rice from 2 to 5 million hectares, to be able to increase rice yield by 1.5 to 2 million tonnes of rice every year, and feed the teeming millions in the next 15 to 20 years. India has 59 hybrid rice varieties released form the public (31 varieties) and private (28 varieties) institutions.

With the proven impact of hybrid seed technology, new tools for hybrid breeding were discovered and utilized for self-pollinating crops including cytoplasmic male sterility (cms). Cytoplasmic male sterility is a condition where the plant is unable to produce functional pollen and would rely on other pollen source to produce seeds. This greatly facilitates large scale hybrid seed production, by-passing hand pollination.

Current hybrid seed technology uses three lines in order to produce the hybrid seed: a) the A line which contains a defective mitochondrial genome in the cytoplasm and a suppressed restorer gene, b) the B line which is genetically similar to the A line but contains a normal cytoplasm and a suppressed restorer gene, and c) the restorer line, a distinctly unrelated line which contains normal cytoplasm and an active restorer gene (dominant).

The two line hybrid system, another hybrid seed technology relies on temperature and geographic location affecting the nuclear genome of the plant, manifested as male sterile. Hybrid seed technology assures hybrid vigor in the progenies but discovery and development of cms lines requires a lot of work and time.

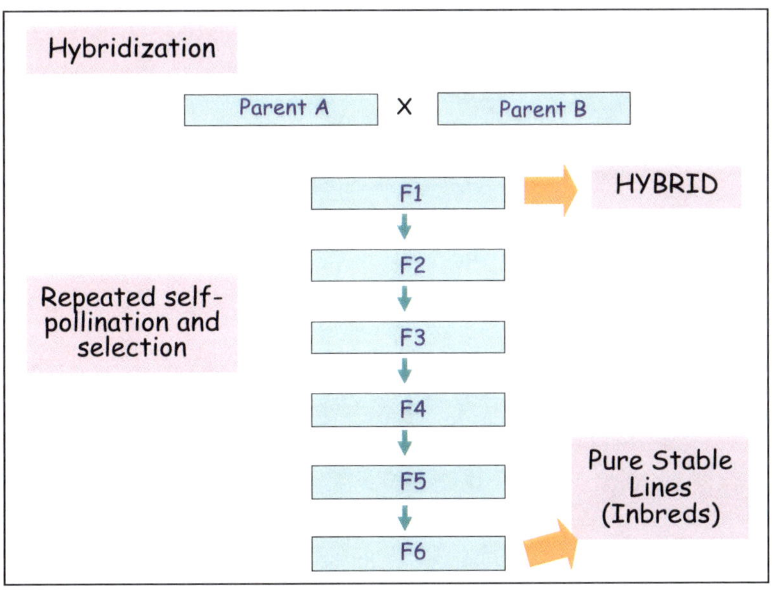

Pure line (inbred line) development.

Conventional plant breeding resulting in open pollinated varieties or hybrid varieties has had a tremendous impact on agricultural productivity over the last decades. While an extremely important tool, conventional plant breeding also has its limitations. First, breeding can only be done between two plants that can sexually mate with each other. This limits the new traits that can be added to those that already exist in that species. Second, when plants are crossed, many traits are transferred along with the trait of interest including traits with undesirable effects on yield potential. Agricultural biotechnology is an option for breeders to overcome these problems.

Tissue Culture and Micropropagation

Plants usually reproduce through sexual means – they have flowers and seeds to create the next generation. Egg cells in the flowers are fertilized by pollen from the stamens (male part) of the flower of the same plant (self-pollination) or another plant (cross). Each of these sexual cells contains genetic material in the form of DNA. During sexual reproduction, DNA from both parents is combined creating offsprings similar to the parents (in self-pollinated crops), or in new and unpredictable ways, creating unique organisms (in cross-pollinated crops). Some plants and trees on the other hand need several years before they flower and set seeds, making plant improvement difficult. Plant scientists have developed the science and art of tissue culture to assist breeders in this task.

Anther Culture of Rice.

Tissue culture is the cultivation of plant cells, tissues, or organs on specially formulated nutrient media. Under the right conditions, an entire plant can be regenerated from a single cell. Plant tissue culture is a technique that has been around for more than 30 years. There are several types of tissue culture depending on the part of the plant (explant) used.

Anther culture is a tissue culture method used to develop improved varieties in a short time. Pollen within an anther contains half dose of the genome (haploid) which spontaneously double (diploid) during culture. In some species however, colchicine treatment is necessary to induce doubling. Doubling of the genome will allow the expression of recessive traits which were suppressed, masked or undetected in routine plant breeding.

Anthers are placed in a special medium, and immature pollen within the anther divide and produce a mass of dividing cells termed as callus. Healthy calli (plural of callus) are picked and placed in another medium to produce shoots and roots (regeneration). Stable plantlets are allowed to grow and mature in the greenhouse. Plant breeders can then select the desired plants from among the regenerated plants.

Anther culture of F1 plants which are progenies in a specific breeding objective would allow many more different types of regenerants. This is because the genetic constitution of the pollen will be more varied than those from the inbreds, thus breeders will have a wider range of traits to choose from. This technology has been employed in the successful development of doubled haploid lines of rice, wheat, sorghum, barley, and other field crops.

Rice varieties developed through anther culture (AC) were released by the National Seed Industry Council of the Philippines since 1995. The first AC-derived, salt tolerant variety PSBRc50 (Bicol) was developed by IRRI and released in 1995. The Philippine Rice Research Institute developed eight salt tolerant varieties and two rainfed varieties.

Micropopagation is a tissue culture method developed for the production of disease-free, high quality planting material and for rapid production of many uniform plants. Actively-dividing young cells (meristem) are placed in a special medium and treated with plant hormones to produce many similar sister plantlets. Since the meristem divides faster than disease-causing virus, clean materials are propagated and hundreds of uniform plantlets are produced in a short time.

Through micropropagation, it is now possible to provide clean and uniform planting materials in plantations – oil palm, plantain, pine, banana, abaca, date, rubber tree; field crops – eggplant, jojoba, pineapple, tomato; root crops – cassava, yam, sweet potato; and many ornamental plants such as orchids and anthuriums. Micropropagated plants were found to establish more quickly, grow more vigorously and taller, have a shorter and more uniform production cycle, and produce higher yields than conventional propagules.

Embryo rescue involves the culture of immature embryos of plants in a special medium to prevent abortion of the young embryo and to support its germination. This is used routinely in breeding parental lines having different or incompatible genome such as in introducing important traits of wild relatives into cultivated crops.

The development of a new rice plant type for West Africa (NERICA – New Rice for Africa) was a result of wide crosses between the Asian Oryza sativa and the African rice Oryza glaberrima. It employs embryo rescue in the initial breeding and in the successive back crossing work followed by anther culture to stabilize the breeding lines. The new plants had combined yield traits of the sativa parent with local adaptation traits from glaberrima.

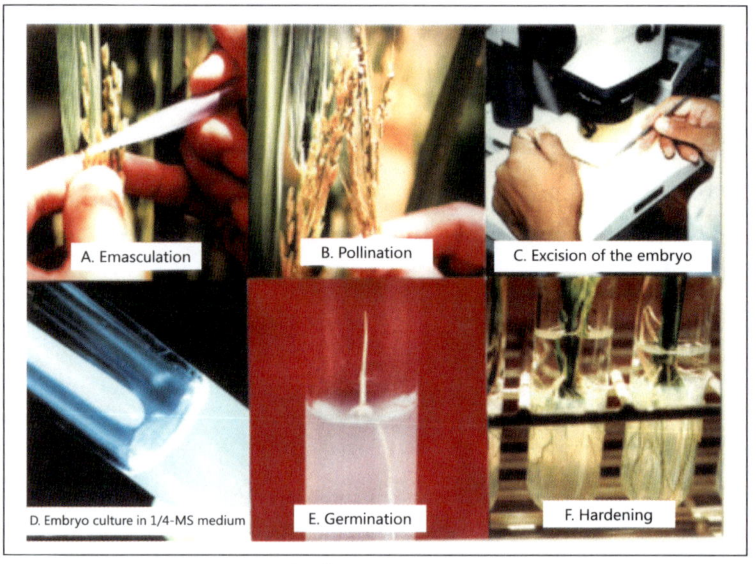

Embryo Rescue.

Wild rices are a rich source of traits for resistance to pests and abiotic stresses. At the International Rice Research Institute, embryo rescue is utilized and facilitated the transfer of bacterial blight resistance genes from wild rice Oryza longistaminata to variety IR24 resulting to a bacterial blight resistant line (IRBB21). Oryza rufipogon is a source of tungro resistance to a number of rice varieties.

At IRRI, a new super salt tolerant rice was developed by saving the embryo produced in the cross between highly salt tolerant wild rice Oryza coarctata with cultivated rice variety IR56. The research team led by Dr. Kshirod Jena has been attempting to cross the two rices since mid 1990s and has only been successful fairly recently. Selected salt tolerant lines will be tested further by farmers in salt affected locations for a possible release within 4 to 5 years.

Plant tissue culture belongs to the lower end of the agricultural biotechnology ladder. But the plant's ability to regenerate a new plant is an important requisite in the development of improved crops through agricultural biotechnology.

Plant tissue culture is a straightforward technique and many developing countries have already mastered it. Its application only requires a sterile workplace, nursery, green house, and trained manpower. Unfortunately, tissue culture is labor intensive, time consuming, and can be costly.

Molecular Breeding and Marker-assisted Selection

The process of developing new crop varieties requires many steps and can take 10 to 25 years depending on the crop. Now, however, applications of agricultural biotechnology have considerably shortened the time it takes to bring them to market. It currently takes 7-10 years for new crop varieties to be developed. One of the tools, which make it easier and faster for scientists to select plant traits is called marker-assisted selection (MAS).

The different traits and physical features of plants are encoded in the plant's genetic material, the deoxyribonucleic acid (DNA). The DNA occurs in pairs of chromosomes (strands of genetic material), one coming from each parent. The genes, which control the plant's characteristics, are specific segments of each chromosome. All of the plant's genes together make up its genome.

Some traits, like flower color, may be controlled by only one gene. Other more complex characteristics, however, like crop yield or starch content, maybe influenced by many genes. Traditionally, plant breeders have selected plants based on their visible or measurable traits, called the phenotype. But, this process can be difficult, slow, influenced by the environment, and costly – not only in the development itself, but also for the economy, as farmers suffer crop losses.

As a shortcut, plant breeders now use molecular marker-assisted selection. To help identify specific genes, scientists use what are called molecular markers which are short strings or sequence of nucleic acid which makes up a segment of DNA. The markers are located near the DNA sequence of the desired gene. Since the markers and the genes are close together on the same chromosome, they tend to stay together as each generation of plants is produced. This is called genetic linkage. This linkage helps scientists to predict whether a plant will have the desired gene. If researchers can find the marker for the gene, it means the gene itself is present.

As scientists learn where each of the markers occurs on a chromosome, and how close it is to a specific gene, they can create a map of the markers and genes on specific chromosomes. This genetic linkage map shows the location of markers and genes, and their distance from other known genes. Scientists can produce detailed maps in only one generation of plant breeding.

Previously, scientists produced very simple genetic maps using conventional techniques. It was observed long ago that as generations of plants were crossed, some traits consistently appeared together in the new generations (genetic linkage). However, since researchers could concentrate on only a few traits in each attempt at cross-breeding, it took many crosses to obtain even a very simple genetic map. Using very detailed genetic maps and better knowledge of the molecular structure of a plant's DNA, researchers can analyze a tiny bit of tissue from a newly germinated seedling. They don't have to wait for the seedling to grow into a mature plant to test for the presence of the specific trait. Once the tissue is analyzed through molecular techniques, scientists know whether that seedling contains the appropriate gene. If it doesn't, they can quickly move on and concentrate analysis on another seedling, eventually working only with the plants which contain the specific trait.

Currently, molecular marker-assisted breeding, an agricultural biotechnology tool is already a routine step in breeding of most crops where the gene and the markers for a specific trait are known. This technique is being used in the efficient introgression of important genes into various crops including bacterial blight resistance in rice, increased beta carotene content in rice, cassava, and banana, and submergence tolerance in rice, to name a few.

Molecular markers are also used to determine the genetic profile of a line or variety. Random primers are used to scan the genomic constitution of the plant through molecular methods. The information is fed to a computer program that will analyze the relatedness of one line to another. The information on genetic diversity of the lines is utilized in selecting for extremely unrelated parents useful for hybrid seed technology. The information will also provide details on the parentage of the line, the possible traits, and the unique identity of the plant useful for germplasm collection database.

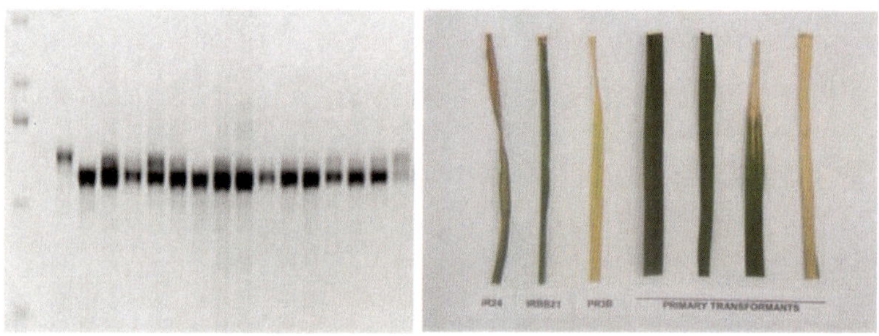

Molecular marker-assisted breeding.

Increasing selection efficiency by selecting for markers associated/linked with the trait of interest.

It should be noted, however, that molecular breeding through marker assisted selection is somewhat limited in scope compared to genetic engineering or modification because: 1) it only works for traits already present in a crop; 2) it cannot be used effectively to breed crops which have long generation time (e.g. citrus); and 3) it cannot be used effectively with crops which are clonally propagated because they are sterile or their offsprings does not resemble the parents. This includes many staples such as yams, bananas, plantain, sweet potato, and cassava.

Genetic Engineering and GM Crops

Over the last 30 years, the field of agricultural biotechnology has developed rapidly due to the greater understanding of DNA as the chemical double-helix code from which genes are made. Genetic engineering is one of the modern agricultural biotechnology tools that is based on recombinant DNA technology. The term genetic engineering, often interchanged with terms such as gene technology, genetic modification, or gene manipulation, is used to describe the process by which the genetic makeup of an organism can be altered using "recombinant DNA technology." This involves the use of laboratory tools and specific enzymes to cut out, insert, and alter pieces of DNA that contain one or more genes of interest. The ability to manipulate individual genes and to transfer genes between species that would not readily interbreed is what distinguishes genetic engineering from traditional plant breeding.

With conventional plant breeding, there is little or no guarantee of obtaining any particular gene combination from the millions of crosses generated. Undesirable genes can be transferred along with desirable genes or while one desirable gene is gained, another is lost because the genes of both parents are mixed together and re-assorted more or less randomly in the offspring. These problems limit the improvements that plant breeders can achieve, eating time and funds along the way.

In contrast, genetic engineering allows the direct transfer of one or just a few genes, between either closely or distantly related organisms. Not all genetic engineering techniques involve inserting DNA from other organisms. Plants may also be modified by removing or switching off particular genes and genetic controls (promoters).

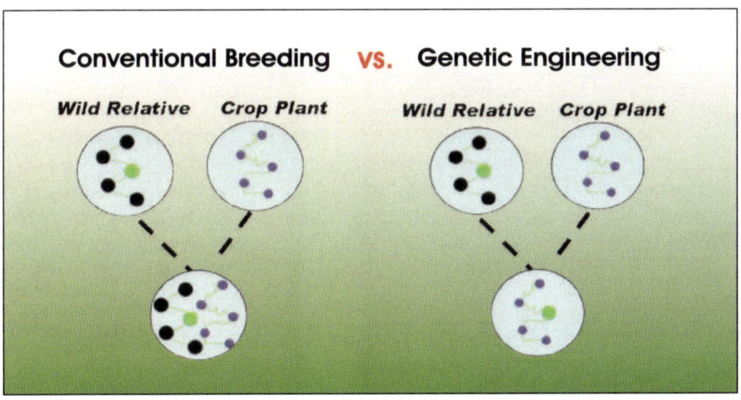

Conventional vs. genetic engineering.

Application of Genetic Engineering in Crop Production

Genetic engineering techniques are only used when all other techniques have been exhausted and when: 1) the trait to be introduced is not present in the germplasm of the crop; 2) the trait is very difficult to improve by conventional breeding methods; and 3) it will take a very long time to introduce and/or improve such trait in the crop by conventional breeding methods.

Modern plant breeding is a multi-disciplinary and coordinated process where a large number of tools and elements of conventional breeding techniques, bioinformatics, biochemistry, molecular genetics, molecular biology and genetic engineering are utilized and integrated.

Development of Transgenic Crops

Although there are many diverse and complex techniques involved in genetic engineering, its basic principles are reasonably simple. It is however, very important to know the biochemical and physiological mechanisms of action, regulation of gene expression and safety of gene and gene product to be utilized.

The process of genetic engineering requires the successful completion of a series of six steps.

Step 1. Nucleic acid (DNA/RNA) Extraction:

Nucleic acid extraction, either DNA or ribonucleic acid (RNA) is the first step in the genetic engineering process. It is therefore important that reliable methods are available for isolating these components from the cell. In any isolation procedure, the initial step is the disruption of the cell of the desired organism, which may be viral, bacterial or plant cells, in order to extract the nucleic acid. After a series of chemical and biochemical steps, the extracted nucleic acid can be precipitated to form thread-like pellets of DNA/RNA.

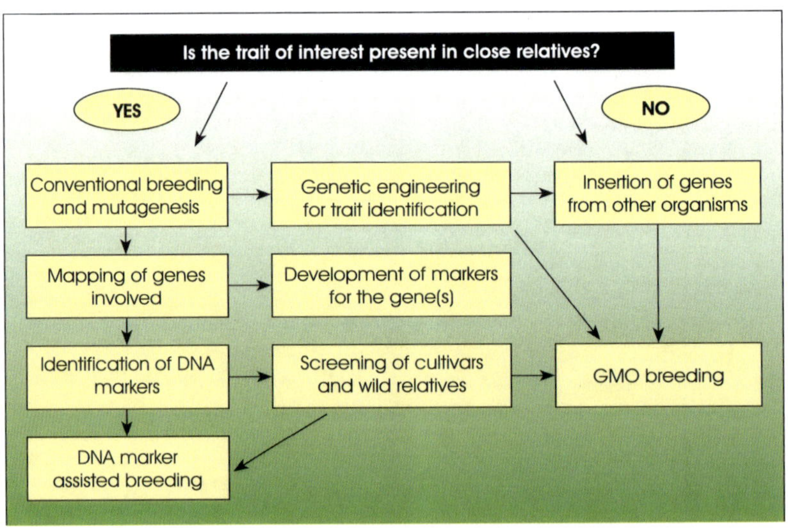

Integration of conventional and modern biotechnology methods in crop breeding.

Step 2. Gene cloning:

The second step is gene cloning. There are basically four stages in any cloning experiment: generation of DNA fragments, joining to a vector, propagation in a host cell, and selection of the required sequence. In DNA extraction, all DNA from the desired organism is extracted. This genomic DNA is treated with specific enzymes called restriction enzymes cutting it into smaller fragments with defined ends to allow it to be cloned into bacterial vectors. Copies of the vector will then harbor many different inserts of the genome. These vectors are transformed into bacterial cells and thousands of copies are produced.

Using information relating to specific molecular marker sequences and the desired phenotype, the vector harboring the desired sequence is detected, selected, isolated and clones are produced. Restriction enzymes are again utilized to determine if the desired gene insert was cloned completely and correctly.

Step 3. Gene Design and Packaging:

Once the gene of interest has been cloned, it has to be linked to pieces of DNA that will control its expression inside the plant cell. These pieces of DNA will switch on (promoter) and off (terminator) the expression of the gene inserted. Gene designing/packaging can be done by replacing an existing promoter with a new one, incorporating a selectable marker gene and reporter gene, adding gene enhancer fragments, introns, and organelle-localizing sequences, among others.

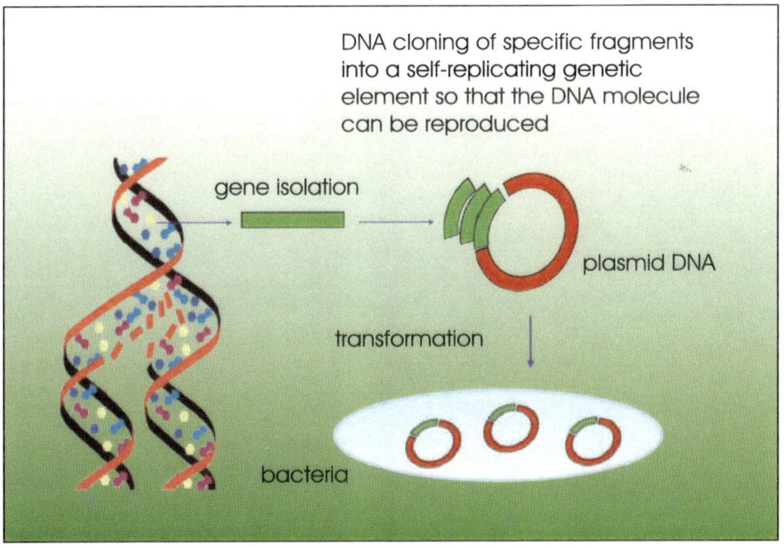

Gene cloning.

Promoters

Promoters allow differential expression of genes. For instance some promoters cause the inserted genes to be expressed all the time, in all parts of the plant (constitutive) whereas others allow expression only at certain stages of plant growth, in certain plant tissues, or in response to external environmental signals. The amount of the gene product to be expressed is also controlled by the promoter. Some promoters are weak, whereas others are strong. Controlling the gene expression is an advantage in developing GM plants.

Selectable Marker Genes

Selectable marker genes are usually linked to the gene of interest to facilitate its detection once inside the plant tissues. This enables the selection of cells that have been successfully incorporated with the gene of interest, thus saving considerable expense and effort.

Genetic engineers used antibiotic resistance and herbicide resistance marker genes to detect cells that contain the inserted gene. Cells that survive the addition of marker agents to the growth medium indicate the presence of the inserted gene. Although increase in antibiotic resistance in humans and animals is unlikely to occur using antibiotic resistance marker, genes coding for resistance to non-medically important antibiotics are preferred. In addition, alternative types of marker genes have been developed which are related to plant metabolism such as phosphomannose isomerase, xylose isomerase and others.

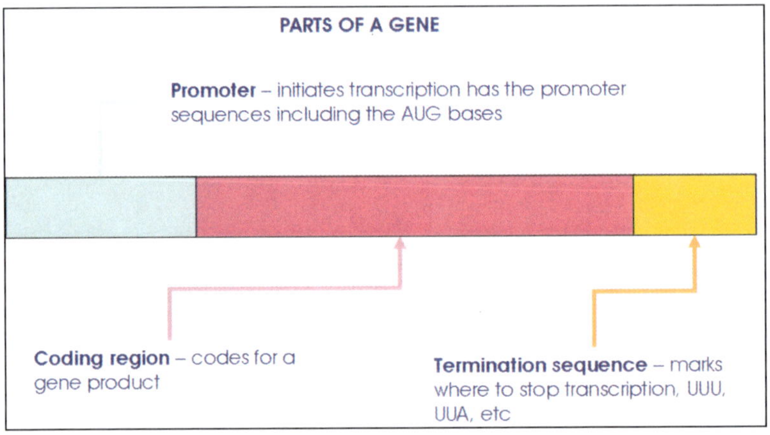

Parts of a gene.

Reporter Genes

Reporter genes are cloned into the vector in close proximity to the gene of interest, to facilitate the identification of transformed cells as well as to determine the correct expression of the inserted gene. Reporter genes that have been used include: the beta glucuronidase gene (gusA gene) which acts on a particular substrate producing a blue product, hence making the transformed cells blue; the green fluorescent protein (gfp) which allows transformed cells to glow under a green light; and luciferase gene that allows cells to glow in the dark, among others.

Enhancers

Several genetic sequences can also be cloned in front of the promoter sequences (enhancers) or within the genetic sequence itself (introns, or non-coding sequences) to promote gene expression. An example is the cloning of the cauliflower mosaic virus promoter enhancers in front of the plant promoter.

Once the gene of interest is packaged together with the promoter, reporter and the marker gene, it is then introduced into a bacterium to allow for the creation of many copies of the gene package. The DNA isolated from the bacterial clones can then be used for plant cell transformation using particle bombardment. If however the use of bacteria Agrobacterium tumefaciens is preferred in the plant transformation, the whole gene package should be cloned in between two border sequences (left and right border) of a binary vector plasmid. This will allow processing of the Agrobacterium so that only the transfer DNA (T-DNA) will be incorporated into the plant genome.

Step 4. Transformation:

The most common methods used to introduce the gene package into the plant cells in a process called transformation or gene insertion, include biolistic transformation using the gene gun and Agrobacterium-mediated transformation.

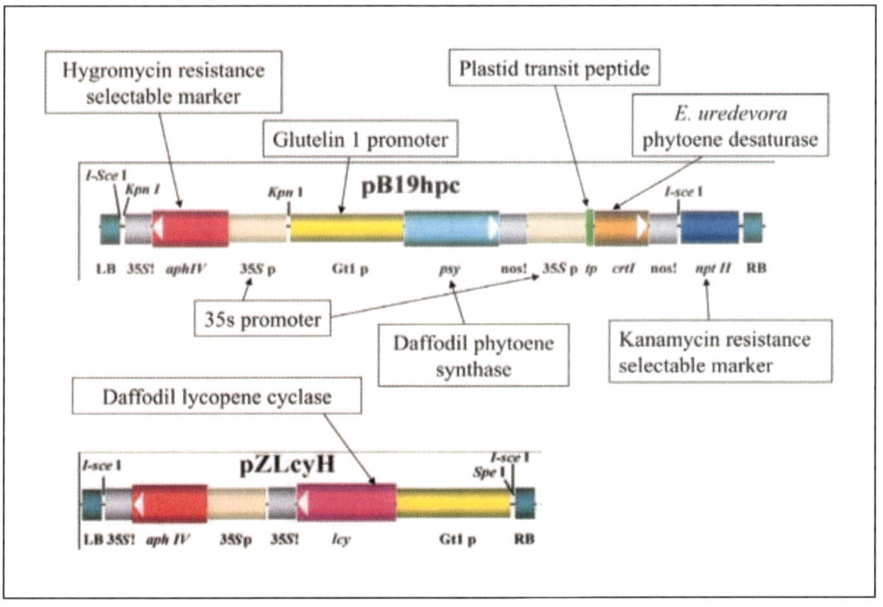

Components of a gene construct used in developing Golden Rice.

Particle Bombardment

Particle bombardment is a mechanical method of introducing the desired gene. The desired genetic sequence is cloned into a plant DNA vector and introduced into the plant using the gene gun or particle gun. As in the common gun, the gene gun uses minute particles of tungsten or gold as the bullet. These particles are coated with the DNA solution and fired to the plant cells through the force of the Helium gas inside a vacuum-filled chamber. The DNA and the tungsten/gold particles get inside the cell, and within 12 hours, the inserted DNA gets inside the nucleus and integrated with the plant DNA. The tungsten/gold particles are sequestered to the vacuole and eliminated later.

Transformed cells are cultured in vitro and induced to form small plants (regeneration) that express the inserted gene.

Agrobacterium Tumefaciens-mediated Transformation

The "sharing" of DNA among living forms is well documented as a natural phenomenon. For thousands of years, genes have moved from one organism to another. For example, Agrobacterium tumefaciens, a soil bacterium known as 'nature's own genetic engineer', has the natural ability to genetically engineer plants. It causes crown gall disease in a wide range of broad-leaved plants, such as apple, pear, peach, cherry, almond, raspberry and roses. The disease gains its name from the large tumor-like swellings (galls) that typically occur at the crown of the plant, just above soil level. Basically, the bacterium transfers part of its DNA to the plant, and this DNA integrates into the plant's genome, causing the production of tumors and associated changes in plant metabolism.

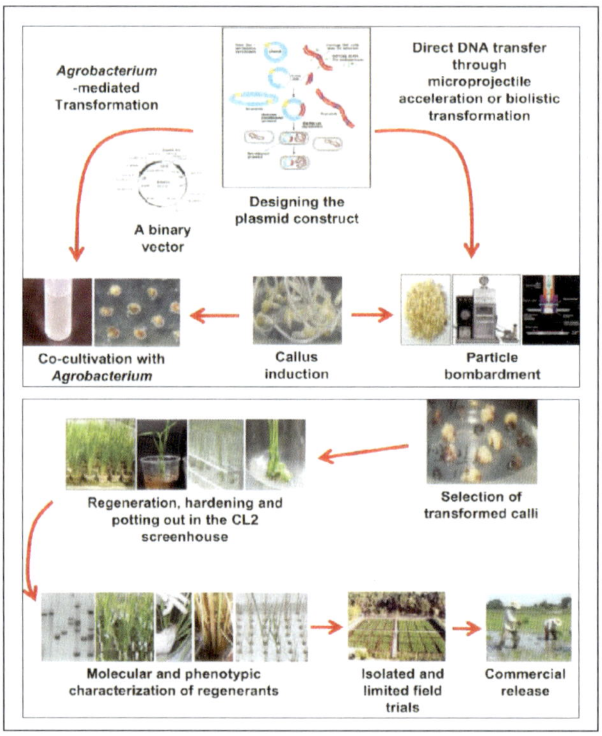

Genetic transformation methods (Biolistics or Gene Gun and Agrobacterium tumefaciens-mediated transformation methods).

Molecular biologists have utilized this biological mechanism to improve crops. The genes that cause the galls are removed and replaced with genes coding for desirable traits. Plant cells infected with the bacterium will not form galls but produce cells containing the desired gene, which when cultured in a special medium will regenerate into plants and manifest the desired trait.

The main goal in any transformation procedure is to introduce the gene of interest into the nucleus of the cell without affecting the cell's ability to survive. If the introduced gene is functional, and the gene product is synthesized, then the plant is said to be transformed. Once the inserted gene is stable, inherited and expressed in subsequent generations, then the plant is considered a transgenic.

Step 5. Detection of Inserted Genes:

Molecular detection methods have been developed to determine the integrity of the transgene (introduced gene) into the plant cell. Polymerase chain reaction or PCR is a quick test to determine if the regenerated transgenic cells or plants contain the gene. It uses a set of primers (DNA fragments) – forward and backward primers, whose nucleotide sequences are based on the sequence of the inserted gene. The primers and single nucleotides are incubated with the single stranded genomic DNA and several cycles of DNA amplification is conducted in a PCR machine. Analysis of the PCR products in agarose gel will show if the plants are really transformed when DNA fragments equivalent in size with the inserted gene is present and amplified.

Southern blot analysis determines the integrity of the inserted gene: whether the gene is complete and not fragmented, at the correct orientation, and with one copy number. The DNA coding sequence is the probe binding to the single stranded genomic DNA of the transgenic plant which is implanted on a nitrocellulose paper. Autoradiography will reveal the transgenic status of the plant. One copy of the transgene is desired for optimum expression.

Northern blot analysis determines whether the transcript or the messenger RNA (mRNA) of the introduced DNA is present and is correctly transcribed in the transgenic plant. The messenger RNA of the transgenic plants are isolated and processed to bind to the nitrocellulose membrane. Labeled DNA is used to bind to the mRNA and can be visualized through autoradiography.

Western blot analysis or protein immuno blotting is an analytical technique used to detect whether the transgenic plants produce the specific protein product of the introduced gene. Protein samples are extracted from the transgenic plants, processed into denatured proteins and transferred to a nitrocellulose membrane.

The protein is then probed or detected using the antibodies specific to the target protein.

Step 6. Backcross Breeding (if needed):

Genetic transformation is usually conducted in elite or commercial varieties which already possess the desired agronomic traits but lacks the important trait of the transgene. Thus, once successfully conducted, the genetically modified plant will be easily recommended for commercialization if it shows stability in several generations and upon successfully passing and fulfilling varietal registration requirements.

However, some plant transformations may have been performed in plant varieties which are amenable to genetic transformation but are not adapted nor important in the target country. There may also be sterility problems in the transgenic plant. In such cases, conventional plant breeding is performed where the transgenic plant becomes the pollen source in the breeding program and the elite lines or commercial varieties as the recurrent parent. Backcross breeding enables the combination of the desired traits of the recurrent parent and the transgenic line in the offsprings.

The length of time in developing transgenic plant depends upon the gene, crop species, available resources and regulatory approval. It varies from 6 to 15 years before a new transgenic plant or hybrid is ready for commercial release.

Commercially Available Crops Improved through Genetic Engineering

With genetic engineering, more than one trait can be incorporated into a plant and are called stacked traits. These are currently corn, cotton, and soybean crops with both herbicide and insect tolerance traits. Transgenic crops with combined traits are also available commercially such as the herbicide tolerant and insect resistant maize and cotton. Stacking different genes for one trait makes the crop more durable to resist the pest/disease and tolerate more herbicides.

Another strategy to improve sustainability of the technology is the use of the refuge. Technology developers have studied effective refuge systems for specific transformation event. These are discussed to farmers extensively for proper implementation, and are monitored regularly to observe any resistant insects or weeds.

New and Future Initiatives in Crop Genetic Engineering

To date, commercial GM crops have delivered benefits in crop production, but there are also a number of products in the pipeline which will make more direct contributions to food quality, clean environment, pharmaceutical production, and livestock feeds. Examples of these products include: rice with higher levels of iron and beta carotene (an important micronutrient which is converted to vitamin A in the body); long life banana that ripens faster on the tree and can therefore be harvested earlier; maize with improved feed value; delayed ripening papaya; papaya ringspot virus resistant papaya; tomatoes with high levels of flavonols, which are powerful antioxidants; drought tolerant maize and wheat; maize with improved phosphorus availability; arsenic-tolerant plants; insect resistant eggplant and rice; edible vaccines from fruit and vegetables; low lignin trees for paper making among others.

Crop Modification Techniques

Traditional Breeding

The most primitive form of plant breeding was the selection of naturally occurring variants in the wild and, later, in cultivated fields. Genetic variation was continuously submitted to the selection pressure of food gathering or planting–harvesting cycles. In some cases, this process resulted in deep changes in plant phenotypes, as exemplified by the derivation of maize from teosinte. This early phase of plant breeding spans the period from the origin of agriculture until the first hybridization experiments carried out by Kölreuter in the 1760s. With the discovery of the laws of heredity, in the turn from the 19th to the 20th century, the importance of hybridization in plant breeding became widely recognized. Today, almost all plant breeding programs involve some use of hybridization.

Plant Domestication

For a given gene, mutations are rare events, but considering the large numbers of plants in a field and of genes in a plant, mutations are quite frequent events in a population. Most mutations are unfavorable for survival in the wild, being eliminated from the population in a few generations, as a consequence of natural selection. However, some of these mutations may result in more favorable phenotypes either in terms of cultivation or in terms of food quality. Some of those mutants were rescued by ancient farmers, who protected them against competition and established with those otherwise disabled plants a relationship of symbiosis. Unlike wild habitats, cultivated fields were environments in which those mutations conferred a selective advantage, thus becoming the predominant type through human selection. The accumulation of this type of mutation is the major cause of the domestication syndrome, a set of characteristics that made many cultivated species irreversibly dependent on humans for their survival.

The molecular variability in domesticated plants tends to be smaller than in related wild species, as a consequence of the *founder effect* during domestication. By strongly selecting for the rare mutant plants adapted to cultivation, early farmers dropped most of the variation present in the wild populations from which cultivated forms arose. It is now clear that many valuable genes, especially those related to resistance to pests, were left out of the cultivated gene pool. Incorporating those genes into modern cultivars, without losing ground in terms of yield and product quality, is one of the challenges of modern plant breeding and one of the most relevant applications of molecular tools in breeding programs.

Intuitive Farmer Selection

Landraces are populations of plants that have been cultivated for many generations in a certain region, being shaped by biotic and abiotic stresses, crop management, seed handling, and eating preferences. They are dynamic genetic entities: continuously changing as a consequence of intentional and unintentional selection, seed mixture, and pollen exchange. Landraces are shaped by a balance between *stabilizing selection*, which keeps the identity of the landrace in a given region, and mild *directional selection*, leading to slow adjustments to environmental changes. In some cases, quick changes can take place, especially when the landrace is taken to a different region or when new materials are cultivated in close proximity with the original landrace. Landraces can still nowadays derive from modern cultivars, if certified seed production is discontinued and farmer-saved seeds are planted recurrently, without care for isolation against seed or pollen contamination.

The major characteristics of landraces are (i) high levels of genetic diversity within populations, characterized by a limited range of variation between individuals, with distinctive traits that make the landrace identifiable; (ii) adaptation to soil and climate conditions typical of the region, combined with resistance to common pests; (iii) edible parts that are valued by local people, normally shaping and being shaped by the local cuisine; and (iv) modest but stable yield, conferring food security to the local community under normal environmental variation.

Intuitive farmer selection has the virtue of shaping varieties for the actual and specific environment of use and for the local food preferences, serving well the case of subsistence agriculture, where most of the production is locally consumed. However, when farmers select for one trait, genetic correlations may result in undesirable changes in other traits. For example, cereal landraces are normally tall plants, prone to lodging and presenting low *harvest index*, probably as a result of human selection for large edible parts (panicles, ears, spikes).

Nevertheless, for their wealth in genetic variability and adaptability to different environments, landraces are the most valuable genetic resources for long-term plant breeding programs and also prime targets for germplasm collections. Seed banks around the world keep thousands of samples of landraces under “ex situ” conservation. In some countries, there are efforts to devise regulatory mechanisms and financial incentives for traditional communities to keep growing their heritage varieties, aiming for their “in situ” conservation. New systems of germplasm conservation have been built on social networks connecting people interested in the subject as a hobby. Those networks take advantage of modern communication tools to replicate on a global scale what used to happen through personal contact in traditional communities.

Notwithstanding, a large part of the variability that once existed in cultivated fields of annual plants may have been irreversibly lost during the introduction of modern, high-yielding cultivars. In this sense, the same modern cultivars that saved millions from starving may have wiped out varieties that were the result of centuries of local intuitive selection by farmers and a valuable resource for future genetic improvement.

Pure Line Selection and Mass Selection

The earliest method of plant breeding based on an elementary knowledge of the laws of inheritance has been the selection of plants within landraces, based on the assumption that the progenies of the best individuals are expected to be superior to the progeny of a random sample of the population. This method was formally proposed by Louis de Vilmorin in 1856, although there are mentions of the use of its principles by some farmers earlier in the 19th century. This realization can be considered as the origin of the paradigm of homogeneity that dominates breeding, and agriculture as a whole, until today. From this point on, within-field heterogeneity was considered to be undesirable and both plant breeding and agronomy developed methods to achieve maximum spatial homogeneity (e.g., "precision agriculture").

In self-pollinating species, such as rice and wheat, landraces can be thought of as a mixture of pure lines, including some heterozygous individuals derived from a low frequency of cross-pollination. In this type of population, selecting single plants and deriving inbred progenies invariably result in some lines that outperform the original landrace for a given growing condition. However, this superiority comes at a cost, because pure lines are normally less stable than diverse populations in the face of stresses, especially diseases, and have no capacity for long-term adaptation, because it is monomorphic for most genes.

In the case of open-pollinated species, such as maize, landraces are populations of random mating individuals, approximating the *Hardy–Weinberg equilibrium* (HWE), with some deviations due to mild selection. Mass selection and recombination of the top-performing portion of the population result in a gradual increase in the frequency of favorable alleles. Successive generations of selection on maize landraces resulted in improved open-pollinated varieties, which were the basis of corn production until the advent of hybrid maize.

Plant Breeding based on Controlled Mating

Despite the great spontaneous diversity that can be found in the landraces, simply applying selection on preexisting diversity is an eroding process that eventually comes to a limit. The true creative power of plant breeding resides in promoting recombination for shuffling favorable alleles. The combination of different alleles in many loci results in a virtually infinite number of genotypes. One could conceivably start a commercial breeding program from a dozen well-adapted founding parents, with a clear focus on a specific target environment and evaluating large segregating progenies. Injection of novel variability might become necessary in the case of a significant change of the target environment, such as the emergence of new pests for which the founder materials had no resistance.

Given the myriad of possible genotypes resulting from crossing diverse parents, the limitation for genetic gains becomes the capacity of the breeding program to evaluate a large number of plants, derived from a large number of crosses. For this reason, plant breeding is frequently dubbed a numbers game, and large competitive programs in commodities invest heavily in high-throughput methods for seed handling, planting, evaluating, and harvesting. As genetic gains accumulate, the bar is gradually raised, and increasingly higher investments are required to keep a steady rate of genetic progress. The limit of this escalation is the financial viability of returns in the seed market and associated business.

Pedigree Breeding: Playing with Parents

The vast majority of the released cultivars of self-pollinating species have been developed through the pedigree method. Pedigree breeding consists of crossing parents and generating segregating populations, which are conducted through generations of self-pollination and selection, until a set of derived lines that combines the good characteristics of both parents is obtained. Because it is based on the complementation of traits, this method is efficient for breeding for qualitative traits, such as disease resistance, or easily classifiable traits, such as plant architecture or the color or shape of plant parts. The pedigree method is appealing to breeders because it allows building better varieties by putting together, in the same plant, good characteristics that were present in different materials. Because all crossings are controlled, it is possible to know the genealogy of each cultivar.

The main weakness of the pedigree method resides in the fact that yield is evaluated efficiently only at the end of the process, on inbred lines, when seed is available for replicated trials. At this point, though, unless a large number of lines have been advanced, there is little room for improvement of yield potential. Consequently, the rate of progress for yield resulting from the pedigree method is normally modest, rarely exceeding 1% per year.

Ideotype Breeding: Playing with Traits

The ideotype breeding approach can be regarded as a strategy to improve the capacity of the pedigree method to promote gains for *quantitative traits*, especially yield. It is based on the hypothesis that one can improve complex traits by changing simpler traits that are positively correlated with them. The advantage of this method is that, if the underlying hypothesis proves correct, one could promote significant gain for yield, even with a small breeding program, taking the "smart" approach, as opposed to the "numbers game" approach of large-scale breeding programs. Additionally, it is scientifically attractive to breeders, because they have a chance of changing paradigms in their favorite crop. However, it is important to keep in mind that unfavorable genetic correlations can offset the advantage brought by the traits that make up the ideotype.

Changing the ideotype of a crop often requires looking for variation beyond the boundaries of elite germplasm, which is normally of the current plant type. However, using landraces as parents in breeding programs normally results in marked reduction in yield. For this reason, backcross toward the elite materials is necessary to recover a competitive progeny. If the donor of the new trait is a wild relative, problems of sexual compatibility, flower sterility, grain quality, or seed dispersal (i.e., grain shattering) may persist for several generations.

To avoid undesirable effects of direct introduction of exotic materials into elite breeding populations, those materials are normally used first in a phase of prebreeding, when breeders try to break the association between useful and undesirable traits. Once the new trait needed to assemble the ideotype is inserted into an elite background, those vector lines can be transferred to the elite breeding program.

Population Breeding: Playing with Genetic Variance

Although all breeding methods imply the application of selection pressure on a variable population, the term "population breeding" indicates a method designed to improve the phenotypic performance of an intermating population by increasing the frequency of favorable alleles controlling traits of interest. The simplest version of population breeding is the mass selection method applied to cross-pollinated species, in which the improved population is directly used as a cultivar. Later, more sophisticated schemes of population breeding have been designed, providing the framework for the development of the quantitative genetics theory. In modern population breeding, the objective is to increase the value of the population as a source of elite lines. Improving the mean quality of the population, while preserving the variation within it, results in top individuals that outperform previously existing lines. Those lines can be used as cultivars, in the case of self-pollinated species, or as parents of hybrids, in the case of cross-pollinated species.

Population breeding is an open-ended scheme of consecutive rounds of selection and recombination, thus being also known as recurrent selection breeding. Recurrent selection requires an efficient crossing system, which can be a limitation for self-pollinating species. When applied repeatedly on small populations, leads to depletion of genetic variation and slowing of genetic gains. For this reason, a large effective population size must be kept throughout or else *genetic drift* will override the other forces acting on the population, hindering genetic improvement.

However, even moderate population sizes (e.g., 50 plants) appear to be sufficient to avoid rapid depletion of genetic variation. This is probably due to the fact that, whereas selection leads to the loss of diversity in some genes, recombination breaks linkage blocks into smaller pieces, allowing many more combinations of genes than was possible with large DNA segments. This phenomenon releases hidden variation in latter generations down the process of recurrent selection, compensating for the loss of variation due to selection.

Breeders create new populations by intercrossing several lines, chosen as sources of favorable alleles for one or more traits. These synthetic populations under recurrent selection mimic the genetic events that used to take place in the landraces in the hands of traditional farmers, with the difference that the whole process is monitored and controlled, and selection pressure is intensified for faster gains. The rate of genetic gain per unit of time can be increased by speeding up the selection–recombination cycles, by intensifying the selection pressure, by improving the evaluation precision (thus increasing the heritability), or by any combination of those.

The general scheme of population breeding is very flexible, allowing customization to specific needs and objectives of different species and breeding programs. The system can be formatted for rapid, short-term results, normally by applying strong selection pressure on genetically narrow-based populations, or for sustained, long-term results, by applying moderate selection pressure on genetically broad-based populations. Population breeding can also be used as a prebreeding scheme, because its frequent crossing events promote recombination between exotic and elite genomes, purging unfavorable exotic genes from the population.

Hybrid Breeding: Playing with Heterosis

Heterosis is the superiority of hybrid individuals compared to inbred individuals. Within certain limits, the more divergent are the parents, the higher is the heterosis on their offspring. Hybrid vigor decays rapidly through generations of inbreeding, indicating that, whatever is the mechanism underlying heterosis, it is due to the presence of heterozygous loci. For this reason, maize breeding programs nowadays are focused on developing competitive *F1 hybrids*, in which heterozygosis is at its maximum.

Two challenges are present in hybrid breeding programs: (i) the need to improve at least two populations toward agronomic adaptation, while keeping them genetically distant enough to express strong heterosis, and (ii) developing efficient seed production of selected hybrids, such that the cost of seed production does not offset the value of the additional yield resulting from heterosis. In maize, the first problem led to the concept of *heterotic groups*, splitting the elite gene pool into subsets, within which population breeding is applied. The second problem was solved by the improvement of parental lines, resulting in higher yield in seed production fields, and by the use of male sterility genes and mechanical emasculation ("detasselling") to avoid self-pollination.

On the other hand, hybrids present great advantages from the business perspective. The seeds produced by hybrid plants are genetically heterogeneous due to the segregation of thousands of genes. If planted, the resulting crop would present a large variation in agronomic traits, plant architecture, and cycle duration, thus reducing yield and grain quality. Hence, farmers must buy new seeds every year, resulting in a constant demand for hybrid seeds. Additionally, with the advent of the transgenics, hybrid seeds became preferential carriers of those valuable proprietary traits because they allow a better control of the event by the owner.

Plant Breeding based on Monitored Recombination

Traditional breeding methods were based on the complementarity between parental characteristics. However, little or nothing was known about which part of the genome came from each parent. This situation changed with the advent and dissemination of molecular marker technologies, which made it possible to monitor the transmission of chromosome segments in the progeny.

Virtually any sequence variation between individuals can be used to design a marker that will allow the identification of the parent that contributed a specific segment of the chromosome in a recombinant line. Until recently, the most popular markers were the simple sequence repeats (SSRs), also known as microsatellites. Those markers were superseded by the single nucleotide polymorphisms (SNPs), which are more abundant in the genome and more amenable to high-throughput genotyping.

Molecular markers are essential tools for studying the genetic control of any trait of interest, eventually leading to the identification of the genes underlying the trait and the metabolic chains involved. This venue can be broadly defined as molecular biology, which dominated the field of biological sciences in the past decades.

Mapping Genes of Interest

When many molecular markers are genotyped in a set of plants derived from a single cross, the frequency of recombination between them can be used to infer their order and relative distance in the chromosomes, resulting in a genetic map. If those plants, or their progeny, are evaluated for a quantitative trait, a statistical model can be built in which part of the phenotypic variance can be explained by some of the markers, which implies that those markers should be linked to the genes underlying the trait. This approach results in *quantitative trait loci* (QTL) maps, which are normally the first step toward understanding the genetic control of a quantitative trait. From a QTL map, a plant breeder can get an insight into the number of loci controlling a trait, their relative importance, and their approximate position in the genome. In the case of large-effect QTLs, it may be possible to jump directly into marker-assisted selection, using the closest markers available in the map. However, in most cases a step of fine mapping, involving a larger population and denser marker spacing in the target genomic region, is necessary for developing useful selection tools. The ultimate result of this approach is the identification of the gene, and the polymorphism in its nucleotide sequence, responsible for the observed phenotypic differences.

The QTL mapping approach is effective for explaining contrasts between two parents, but is inefficient for exploring the wider genetic diversity for a trait in the germplasm. The association mapping approach offers a shortcut in this path and for this reason has strong appeal to breeders. In this approach, a panel of genetically diverse lines is densely genotyped and carefully evaluated for a phenotypic character, with the aim of identifying associations between marker alleles and traits. Because the genealogical distance between the materials in the panel is large, those associations will remain significant only if the marker is tightly linked to the causative gene or if factors related to population genetics (population structure) create associations between unlinked loci. For the latter case, estimating population structure and taking it into account explicitly in the statistical model can avoid the detection of false associations. When correctly applied, association analysis can result in the detection of genomic regions related to the trait of interest and, simultaneously, in the identification of donor lines of favorable alleles from a wider germplasm.

Gene banks harbor thousands of accessions that are potentially useful for plant breeding. Those accessions include wild crop relatives and obsolete landraces. Although presenting poor agronomic value, compared to modern cultivars, those materials are believed to have useful genes that have not been captured from wild species in the process of domestication or from landraces in the early phases of scientific breeding. Rescuing those useful genes that were "left behind" is a difficult task, because the genes of interest may be tightly linked to unfavorable genes, which would be dragged along into the breeding population. Understanding the genetic diversity in germplasm collections is the first step toward better use of a broader gene pool in breeding programs. Marker-assisted backcrossing allows identifying the rare recombinants in the vicinity of the introduced gene, trimming the chromosome segment as close as possible to the target gene. Additionally, by identifying individuals with the cleanest genome (with the least residue from the donor parent), it is possible to reduce the number of backcrosses needed for full recovery of the elite phenotype.

Collections of chromosome segment substitution lines (CSSL libraries) are a set of lines derived from an elite variety, in which each line has one chromosome segment replaced by the corresponding segment in the wild species of interest. The CSSL library carries the whole genome of the wild relative split in bins. The phenotypic effect of the set of genes in each bin can be evaluated against the original elite line, used as a check. Those genetic stocks can facilitate the identification and introgression of genes from wild relatives into elite germplasm. Superior CSSLs can be used for fine mapping and simultaneous development of an improved version of the elite variety.

Marker-assisted Selection: Building Tailored Genotypes

Whereas in the case of marker-assisted gene introgression the breeder will normally tag only one gene, a full-featured marker-assisted selection scheme would monitor several genes simultaneously. Knowledge of the allele borne by each plant in each major locus of interest creates the means for building specific allele combinations that would maximize the agronomic value of the line. However, when one is dealing with several genes, it is important to keep in mind that the interaction between genes (epistatic effect), and not only the additive effects of the genes, defines the expression of a trait.

A typical application of marker-assisted selection is resistance gene pyramiding. When one is dealing with diseases caused by pathogens presenting high variability, a single resistance gene may confer complete resistance for a few plant generations until it is overcome by strains of the pathogen. Accumulating several resistance genes in a variety can confer durable resistance, because the pathogen would have to beat all of the genes simultaneously, which drastically reduces the odds of resistance breakdown. Resistance gene pyramiding by classical breeding methods is almost impossible, because the effect of a gene could be visible only in the absence of other resistance genes. Marker-tagged resistance genes can be easily combined without the need for phenotypic screening. A similar strategy can be used for preventively breeding resistance to pathogens that are currently absent from a region, because in the absence of the pathogens no phenotypic screening of resistance would be possible.

Molecular markers of interest for breeders of a given species can be combined in optimized sets for simultaneous genotyping ("multiplexed"), allowing high-throughput application in applied breeding programs, especially those dealing with a broader genetic base. The implementation of such a marker set depends on a large body of previous research. For each of the traits of interest, several steps of research are required, normally starting with mapping quantitative trait loci (QTL), followed by fine mapping and eventually positional cloning. Once the causative gene is identified, based on a pair of contrasting alleles, "mining" novel alleles in diverse germplasm may expand the range of variation of the trait or the toolbox to deal with specific stresses.

Genomic Selection: Speeding up Genetic Progress

Recent advancements in genotyping technology sharply reduced the cost of genotyping, creating the possibility of scoring thousands of markers in populations of plants under selection. The several steps of genetic analysis required to identify the relationship of each of those markers with the phenotype could not keep pace with that. In this scenario, the breeder can have a large amount of genotypic information with unknown relationship with the traits under selection. The genomic selection approach proposes that knowledge of the relationship between specific markers and specific genes is not necessary in the breeding context. Instead, the breeder can use the information available for all markers in a plant to predict its breeding value, without effectively evaluating its phenotype, based on previous statistical models built for those markers using a "training population" for which all genotypes and phenotypes have been scored. Markers are considered as random factors, in the framework of mixed model analysis, because the number of markers used in genomic selection is normally superior to the number of individuals in the training population, such that estimating the effect of each marker would not be possible due to the lack of degrees of freedom. Simulation studies demonstrated that genomic selection can accelerate the genetic progress of quantitative traits in population breeding.

Genomic selection can be seen as a method to maximize the return from the phenotypic evaluation step in the breeding cycle. The information drawn from one round of proper phenotypic evaluation can be propagated to subsequent generations. Once models are built for different environments, the unique marker profile for a given individual can be used to predict its performance in each of these environments, improving considerably the inference made on each individual performance and largely reducing evaluation costs. The potential use of this approach in choosing parents of new breeding populations is also promising. However, the predictability of the phenotype from the marker profile fades out along generations, and new rounds of phenotypic data must be input to update the statistical model. This approach will be especially useful in cases when phenotypic evaluation is difficult, expensive, or conditioned to uncertain events, such as the occurrence of a pest or specific weather condition.

Genomic selection can be used to accelerate breeding for yield potential, which is frequently considered as the most difficult trait for marker-assisted selection. In highly competitive seed industries, such as the cases of maize and soybean, any increase in the rate of genetic gain could make a significant difference in market share. For this reason, large seed companies are investing heavily in high-throughput DNA analysis and bioinformatics for seamless integration of genomic selection in breeding programs. Seeds can be genotyped before planting, such that only preselected individuals will use expensive plots in the field. Full application of genomic selection should happen more easily in large private companies, when compared to small seed companies, public institutions, and universities, considering the amount of investment required for setting and keeping up-to-date the genotyping pipeline, the short turnaround time for DNA analysis, data processing and selection, and the well-trained team required.

Mutation Breeding

Mutation breeding is the purposeful application of mutations in plant breeding. Unlike hybridization and selection, mutation breeding has the advantage of improving a defect in an otherwise elite cultivar, without losing its agronomic and quality characteristics. Mutation breeding is the only straightforward alternative for improving seedless crops. Since the first release of mutant cultivars that resulted from basic mutation research in Europe, mutation breeding has found a niche in plant breeding because of these advantages. Methodologies for mutation induction have been improved in main crops for both physical and chemical mutagens, and selection methodologies for mutant populations have been described. New mutagenic agents such as ion-beam radiation and cosmic rays are being investigated and a hitherto undescribed wide spectrum of mutations has been observed. Nevertheless, alkylating agents and ionizing radiation are still popular. The development of robust *in vitro* techniques for many crop species has contributed to improving the efficiency of mutation breeding. The ability to handle large mutagenized populations in a confined space, faster progeny turnover in vegetatively propagated species and the ability to screen for several biotic and abiotic stress factors in the culture environment make *in vitro* approaches very efficient. Mutant screening has undergone revolutionary changes in the past decade with reverse genetic approaches taking precedence. Therefore, integration of mutation techniques with molecular approaches is providing exciting opportunities for modern plant breeding.

A mutation is a sudden heritable change in the DNA in a living cell, not caused by genetic segregation or genetic recombination. Mutation breeding can be described as the purposeful application of mutations in plant breeding. Plant breeding can be seen as human-guided evolution of crop plants, and genetic variation is its foundation. When desired variation is available in different cultivars of a crop, its breeding objectives may be achieved through cross-breeding. However, when one or more of the parent cultivars possessing the desired characters used in cross-breeding are poorly adapted, it becomes necessary to implement a back-crossing strategy to recover the elite type. Another problem encountered in conventional crossbreeding is the poor combining ability of some parental genotypes. For example, aromatic rice varieties have poor combining ability, and cross-breeding with non-aromatic varieties will lead to a decrease in aroma and quality. Under such circumstances, mutation induction can be of advantage to produce cultivars with desired traits within defined germplasm pools.

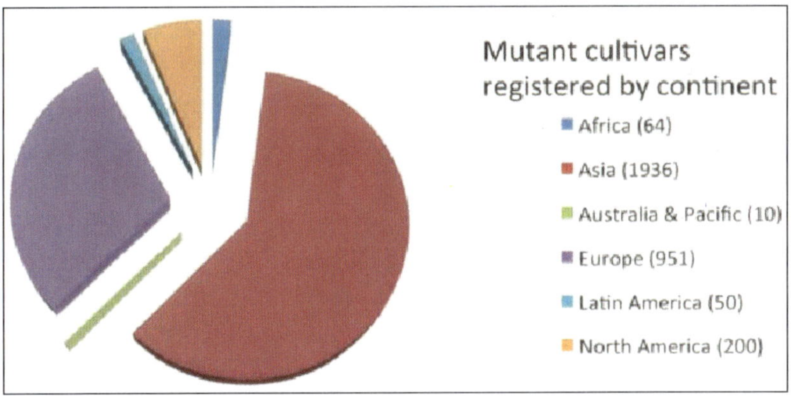

Number and proportion of mutant cultivars released, categorized by continents.

Sometimes genes conferring undesirable characters are tightly linked to genes controlling desirable traits. In such a situation, mutation induction may result in a crossing-over event or isolation of an independent mutation for the desired trait. In crops that do not produce seeds, such as edible banana or seedless grapes, mutation induction may be the only acceptable way of increasing variability for developing new cultivars. This also applies to many root and tuber crops, and the development of novel colours and variations in ornamental plant species propagated vegetatively. Cross-breeding in vegetatively propagated perennial crop species, such as many fruit crops, is also subject to constraints of time, growing space and clonal identity, and mutation induction can be a valuable breeding strategy.

Induced mutations have a history of 83 years since the first publications by Muller and Stadler. Plant mutation breeding has been applied actively for over a half century, with the Joint FAO/IAEA Division based in Vienna assisting developing countries to quickly adopt advances in the technology. This approach helped widespread use of mutation technology and the release of 3211 registered mutant varieties in more than 170 different plant species to date. Some of these mutant cultivars have revolutionized agriculture not only in densely populated developing countries but also in agriculturally advanced countries.

Types of Mutations

Descriptions of different types of mutations, their classification and effects have been reviewed in the literature. Mutations can be broadly divided into intragenic or point mutations (occurring within a gene in the DNA sequence), intergenic or structural mutations within chromosomes (inversions, translocations, duplications and deletions) and mutations leading to changes in the chromosome number (polyploidy, aneuploidy and haploidy). In addition, it is important to distinguish between nuclear and extranuclear or plasmone (mainly chloroplast and mitochondrial) mutations, which are of considerable interest to agriculture.

Intragenic Mutations

Loss or gain of one or more base pairs is referred to as deletions and insertions, respectively, and if they do not occur in multiples of three nucleotides in the sequence, it results in a frame-shift mutation (within exons of coding regions). Therefore, deletion or insertion of a single base pair or two may result in a substantial effect, including the loss of function through a shift within the reading code. However, the effect of such a mutation can depend also on the location where the change happens. Thus, a frame-shift mutation near the 3 end of the gene will result in only a change in the terminal part of the polypeptide chain as translation takes place only in a $5' \rightarrow 3'$ direction. Under such circumstances, even a frame-shift mutation can result in a functionally similar protein.

The other type of gene mutation is base-pair substitution, common with chemical mutagens such as alkylating agents. They result in the incorporation of alternate bases during replication. For example, A-T base pair can be switched to G-C, C-G or T-A. Thus, one base of a triplet codon is substituted by another, resulting in a changed codon. Where that change leads to an altered amino acid sequence, the result is a mutated protein. If a purine base is replaced by another purine base (G by A or A by G) or a pyrimidine by another pyrimidine (T by C or C by T) the substitution is called a transition. Transitions are by far the most common types of mutations and C to T transitions occur more frequently than any other base pair substitutions. If a purine base is substituted by a pyrimidine, or *vice versa*, the substitution is called a transversion. McCallum calculated that in *Arabidopsis*, about 5% of mutations caused by alkylating agent ethylmethane sulfonate (EMS) will introduce a stop codon (non-sense mutation), 65% will be missense mutations (a codon coding for one amino acid is converted to a codon coding for another amino acid), and 30% will be silent changes, where the final protein product remains unchanged.

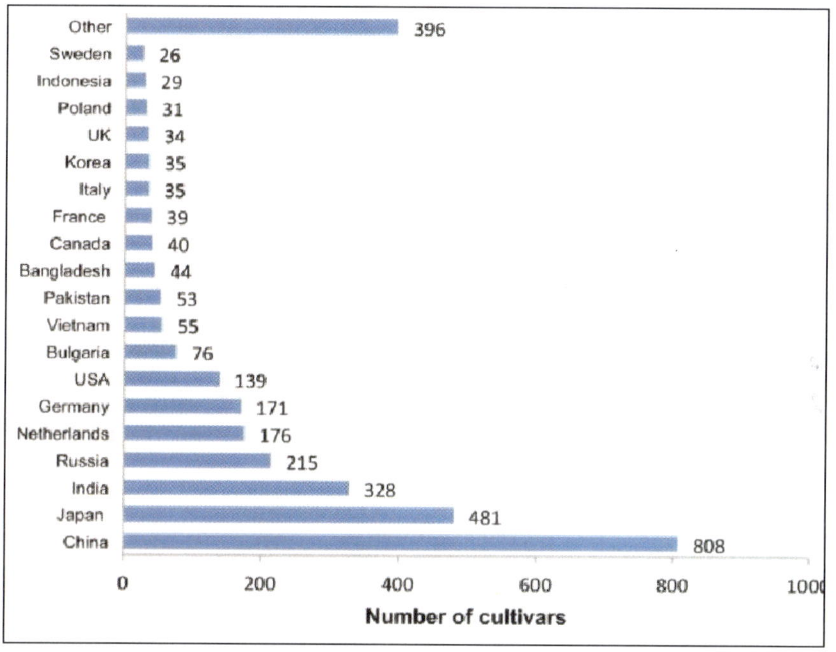

Number of mutant cultivars released in different countries.

Chromosome Rearrangements or Structural Mutations

Structural mutations are the result of chromosome breaks and their rearrangements. Ionizing radiation results mostly in such changes. Four categories of such rearrangements can be distinguished: deletions or deficiencies, duplications, inversions and translocations. About 90% of the chromosome aberrations brought about by ionizing radiation are deletions and often are lethal. However, some deletions can block biochemical pathways and, for example, if this happens in a pathway leading to the synthesis of a toxic metabolite, such a deletion may result in a useful non-toxic plant product. Duplications have been a cornerstone in crop evolution, particularly at diploid level. For example, the smallest known plant genome, that of *Arabidopsis*, carries three major duplications that have taken place between 14.5 and 300 million years ago. Similar evolutionary patterns have been unravelled for sunflower as well. Such studies lead us to the conclusion that most, if not all, angiosperms are palaeopolyploids.

Inversions (180° rotation of blocks of nucleotides) and translocations are also common in evolution of crop plants. For example, by comparative mapping, Heesacker *et al.* identified five translocations and two inversions in cultivated sunflower during its evolution. Emerging evidence of the role of chromosomal rearrangements in crop evolution suggests that crop improvement can be quickened through the combination of distant hybridization coupled with radiation breeding. The first example of this application was demonstrated by Sears by transferring a small section of an *Aegilops umbellulata* chromosome containing a gene for resistance to brown rust (*Puccinia triticina*) to hexaploid wheat through a bridging cross employing *Triticum dicoccoides* and irradiation.

Changes in Chromosome Number – Genome Mutations

Mutations at the genome level, leading to changes in chromosome number are also of utmost importance in crop evolution and plant breeding. These mutations are broadly divided into allopolyploidy, autopolyploidy and aneuploidy. It is estimated that 50–70% of the ornamental flower crops are polyploids. Allopolyploidy is the result of a combination of genomes of two or more species and commonly arises through interspecific or intergeneric hybridization followed by the doubling of chromosomes. Evolution of many crops, such as wheat, soybean, cotton and *Brassica* species, has taken this path. Mutant selection is often complicated by the presence of duplicated homoeologous allele pairs in allopolyploids, but targeting induced local lesions in genomes (TILLING) populations are helpful in identifying and targeting these alleles for breeding. There are examples of crop evolution at a diploid level, and rice, maize, barley, sesame and tomato belong to this group.

Autopolyploidy is the result of multiplication of chromosome number within the species and the nucleus contains more than two homologous chromosomes. Cultivated and tuber producing wild potatoes have a polyploidy series of 24, 36, 48, 60 and 72 chromosomes. Edible banana evolved at a triploid level through polyploidization at a single-genome level as well as through allopolyploidy as a result of interspecific hybridization of the two wild species *Musa acuminata* and *Musa balbisiana*. The great diversity of banana germplasm, comprising fruits suitable for cooking through to soft and sweet desert types, can be attributed to this divergence at a genome level. Modern breeding of sugar beet is based on hybridization of diploids with tetraploids to produce hybrid triploids. Sugar beet produces the highest yield and sugar content at a triploid level. Autopolyploidy can be induced by the use of compounds that inhibit spindle fibre formation during mitosis, thus the divided chromosomes are retained in the same cell. Homozygous lines for heterosis breeding can also be obtained using these chemicals on haploid plants produced through anther, microspore or ovule culture. Colchicine and oryzalin are the most widely used chemicals for artificial polyploidization.

Extranuclear or Plasmone Mutations

In higher plants, two extranuclear genetic systems can be distinguished, those of chloroplast and mitochondria. Mutations in these genomes have been shown to be of value in agriculture. Many authors have shown that nitroso compounds produce a high frequency of plasmone mutations. Using nitroso compounds, chloroplast mutations for antibiotic resistance have been induced in *Nicotiana*. Chlorophyll mutations observed already in the first generation after treatment with mutagens are also the result of such mutations. These mutations can be used to an advantage in breeding ornamental crops with various leaf or fruit characters as they often occur as chimaeras. As plastid mutations are carried through the cytoplasm, they are maternally inherited and do not behave in a Mendelian fashion. Resistance to some fungal diseases and cytoplasmic male sterility are other important characters inherited through plasmonic genes and discussed in detail by van Harten.

Mutagenic Agents and their use in Mutation Breeding

Physical Mutagens

Ionizing radiation was used exclusively in mutation research until the effects of certain chemicals on the DNA were published in the mid-1940s. In the beginning, X-rays were used, but *g* -rays from radioactive sources such as ^{60}Co and ^{137}Cs became popular as they were made available to many developing countries through the IAEA. Fast neutrons from nuclear reactors can also be used, particularly as this irradiation service is available through the Joint FAO/IAEA Division in Vienna. Ionizing radiation results in chromosomal breakages, allowing cross-linking of DNA strands. More deleterious effects, therefore, can be expected than from chemical mutagens that result in minor changes in the DNA such as base-pair substitution. Among physical mutagens, ultraviolet (UV) rays are non-ionizing. They therefore have low penetration and are effective in producing purine or pyrimidine dimers, resulting in point mutations. UV rays can be effectively used to irradiate pollen in the late or early uninucleate stages. Radiation doses and conditions for irradiating seeds and buds have been optimized and published for main crop species. Treatment conditions and doses need to be experimentally derived for new or minor crop species as information may not be readily available. A suitable dose for irradiation has often been recommended as the one that reduces growth and morphogenic performance to 50% (GR 50) of untreated controls, but for seeds it is recommended to use 2–3 doses. The main external factors affecting irradiation are oxygen and moisture contents, temperature and post-irradiation storage conditions.

Chemical Mutagens

High rates of chromosome aberrations resulting from ionizing radiation and the accompanied detrimental effects made researchers look for alternate sources for inducing mutations. As a result an array of chemical mutagens has been discovered. A wide variety of chemical mutagens, however, makes it difficult to establish common rules and conditions for treatment. Classification of chemical mutagens, methods of treatment, post-treatment handling and selection after treatment has been discussed for main crop species in the literature. Since the reporting of mutagenic effects of sodium azide, no new chemical mutagens of widespread use in plant breeding have been discovered. The most widely used chemical mutagens are alkylating agents, with EMS being the most popular because of its effectiveness and ease of handling, especially its detoxification through hydrolysis for disposal. Nitroso compounds are the other alkylating agents widely used, but they are light-sensitive and more precautions need to be taken because of their higher volatility. Chemical mutagens are also popular in *in vitro* mutation induction, although irradiation can also be applied at low doses. EMS has become the mutagen of choice for developing mutant populations for high throughput screening such as in developing TILLING populations.

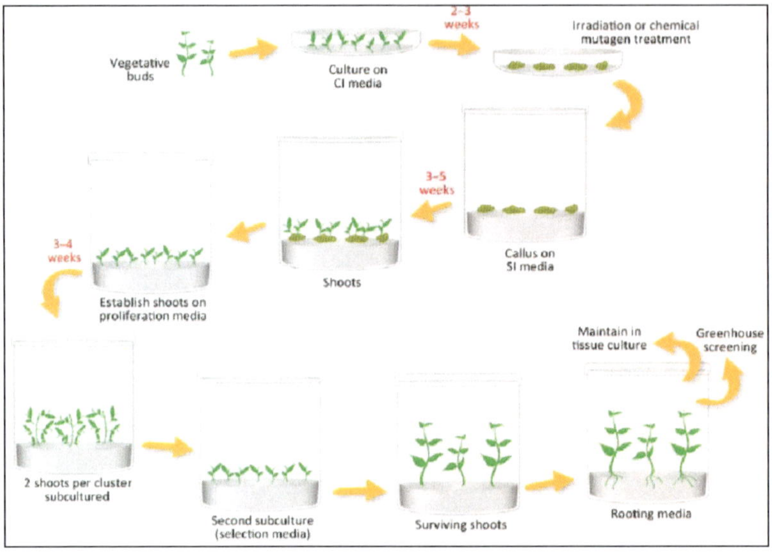

A hypothetical scheme for mutation induction and selection *in vitro*.
CI media – callus induction media, SI media – shoot induction media.

New Methods for Mutation Induction

Since a large proportion of basic research on plant mutations has been accomplished for temperate cereals, most of the straightforward recommendations relate to them. However, this research has been successfully adapted to rice and other tropical cereals, legumes, oilseeds and vegetables as demonstrated by the growing numbers of released cultivars – 3211 registered in over 170 species.

Fast development of *in vitro* based methods in the past two decades has helped plant breeders to develop more efficient induction and screening processes using plant cell and tissue cultures. This has been particularly useful in developing novel variation in vegetatively propagated crops such as banana, cassava, sweet potato, ulluco, sugarcane, *Citrus* sp., Indiangrass and many ornamental species. *In vitro* cultures have the advantage of producing large mutant populations in a limited space, and the possibility of screening for some useful characters such as tolerance to salt, toxic elements or fungal toxins in the laboratory environment. This allows the transfer of only putative mutants for field screening, and in the case of vegetatively propagated crops, removal of chimaera through several subcultures. *In vitro* mutation induction and screening methods have been established for widely cultivated crops such as banana, potato, cassava, sweet potato, grapevine and sugar cane and a hypothetical scheme for *in vitro* mutation induction and selection is given in figure. Callus cultures are treated with the mutagen and individual shoots are regenerated in shoot induction media. Selection presure can be applied at this stage.

Regenerated shoots are then subcultured to remove chimeras in selection media and surviving shoots can be rooted and further tested in the greenhouse before field evaluation.

In recent years, plant mutations have been obtained using high-energy (220 MeV) ion beam implantations in Japan, and in China using low-energy (30 keV) ion beams. Resistance to bacterial leaf blight and blast in rice, yellow mosaic virus in barley, potato virus Y in tobacco and black sigatoka disease in banana are a few examples of useful mutants in field crops developed through ion beam irradiation. Many ornamental crop cultivars with unique colour characteristics have been developed as well. Tanaka *et al.* recently summarized the results of 19 years of work in different crops and concluded that the spectrum of mutations induced by ion beam radiation is different from that of gamma rays, and novel mutant types hitherto not produced have been isolated in ornamental plants. However, in rice the spectrum of different ion beams (C and He) was similar to that of gamma rays, but frequency was higher. Additional information on the use of ion beam mutagenesis can be found in a recent review by Magori *et al.*

Cosmic radiation is another novel source for mutation breeding and a term 'space breeding' has been coined to describe it. Spacecraft, recoverable satellites and high-altitude balloons have been used and more than 60 cultivars have been released as a result of this programme. Both genomic and epigenetic changes have been detected in plants derived from seeds exposed to space flight. Molecular and segregation analysis of three blast-resistant rice mutants induced by space flight revealed that one or two dominant mutations were responsible for resistance to different isolates of the pathogen. Discovery of dominant mutations from space flight is noteworthy and may be indicative of a different spectrum from other mutagenic agents.

Insertional mutagenesis is a useful method for producing mutants to investigate gene function because those mutations are tagged with DNA fragments of known sequence. There are two methods to construct such mutants: T-DNA insertions and transposon tagging. T-DNA insertions are considered genetically modified, carrying T-DNA sequence from *Agrobacterium tumefaciens* and thus pose difficulties associated with regulatory barriers to GM crops in an agricultural environment. On the other hand, transposable elements are an internal genetic factor that has been harnessed recently for mutation breeding through their transcriptional activation. These elements can get incorporated into various regions of the plant genome (insertion mutations) and their position can be identified by PCR-based methods or *in silico* screens of databases that contain the DNA sequences flanking all insertions. Transgenic copies of the maize Activator/Dissociation (Ac/Ds) transposon system has become a very popular tool for gene tagging and functional genomics in various plant species and thus for 'targeted mutations'. Within the last decade populations carrying such insertional mutations have been developed not only for the model plants Arabidopsis and *Lotus japonicus* but also for economically important food crops such as rice, barley, maize and soybean. High density of mutations in these populations has helped researchers to move beyond functional genomics and use them in crop improvement programmes. Jiang and Ramachandran have reviewed the different mutant rice populations available for genetic analysis and estimated that the rice genome encodes about 32, 000 protein coding genes.

Restriction endonucleases have been successfully used to induce double-strand breaks, resulting in mutations. Further perfection of technology has now seen the use of zinc finger nucleases (zinc finger DNA-binding proteins) for inducing site-specific mutations.

Mutation Rates and Detection of Mutations

Mutation Rates

The random nature of mutational events applies not only to natural mutations but also to those induced. Thus, there is little control over the magnitude or kind of genetic change expected. However, there are reasonable estimates on the frequency of mutations. Provided an effective treatment is given, a particular gene could be expected to mutate once in about 10 000 treated cells. Molecular-based methods, particularly reverse genetic approaches, have enabled more accurate estimation of mutation rates in recent years. Thus, TILLING of an EMS-generated *Arabidopsis* mutant population of 3000 plants revealed that 14 base mutations per 1.5 kb fragment length can be discovered. Mutation rates for hexaploid wheat has been estimated at 1 per 25 kb, in *Brassica rapa* 1 per 60 kb, but for maize and barley the rates were as low as 1 per 500 kb. In peanut, the mutation rate for the major allergen gene *Ara h 1* in an EMS population was estimated to be 1 per 967 kb. Both EMS and a combined treatment of sodium azide and *N*-nitroso-*N*-methylurea (NMU) produced about one mutation per < 300 kb length in rice. In two EMS-derived mutant populations of tobacco, Julio *et al.* estimated mutation rates from 1/30 to 1/83 kb for two genes controlling nornicotine biosynthesis. These mutation rates are average figures and estimates, and can vary depending on the gene, kind of mutagen used, plant part used for mutagen treatment, conditions of treatment, post-treatment handling and even the location of cells, for example whether dormant or actively dividing meristematic cells. For example, Cooper *et al.* compared mutation rates in four soybean TILLING populations derived from EMS and NMU. In the NMU and one of the EMS populations the rate was estimated at 1 per 140 kb. In the other two EMS populations the rates were 1 per 250 and 1 per 550 kb. These different mutation rates can serve as guidelines to estimate the number of mutagenized plants that need to be screened in a practical mutation breeding programme.

Detection of Mutations

For practical plant breeding, mutations occurring in those cells that give rise to the progeny are relevant. In the early phases of mutation breeding, most of the selected mutants that led to released cultivars resulted from easily recognizable characters such as early flowering and maturity, changed plant architecture and height, fruit and seed characteristics or resistance to diseases that can be screened easily due to their epiphytotic nature. Development of low-cost, high throughput screening methods resulted in the development and release of mutant cultivars with improved or changed quality, as in the development of linola or the black spot disease resistant mutants of Japanese pear described in the section on the application of mutation breeding and impact of mutant cultivars. *In vitro* screening methods can be applied for traits that can be mimicked in culture such as drought, salinity, toxic elements and for disease resistance, for example using culture filtrates from toxin-producing pathogenic micro-organisms, fungal or bacterial toxins or their analogues. Detection of mutants in haploids is faster (can be detected in the first generation) and easier, and will be directly applicable if the breeding programme is based on doubled haploids as demonstrated by Carlson.

In the last decade new molecular biological tools have been developed that can screen mutants at the gene level. Although genome sequencing is considered the 'gold standard' for mutation detection as it can reveal the exact location of a mutation and its type, applying this to large populations is costly and is rarely used in practical mutation breeding. Moreover, direct sequencing may not readily detect heterozygous alleles. Therefore, several methods to scan amplicons in mutant populations for the presence of sequence polymorphisms have been developed. Among the novel techniques, TILLING has emerged as a robust, high-throughput method that can be applied to many species. The potential of this method in plant agriculture has been reviewed in this series. Since that review, TILLING populations for oat, sorghum, soybean, pea, *B. rapa*, *Brassica napus*, tomato and rice including one of the 'green revolution' cultivars 'IR 64' have been developed and many more populations are in the pipeline. TILLING is a reverse genetics approach and relies on the detection of sequence alterations in target genes to identify allelic variations among mutant populations. TILLING populations are developed by chemical mutagenesis covering the entire genome and then using an enzymatic detection method in combination with gel electrophoresis or denaturing high-performance liquid chromatography. It relies on the detection of single-base mismatches in heteroduplexes using endonucleases. Many TILLING populations have been developed based on a method described by Colbert *et al.*, which relies on a mismatch-specific endonuclease found in celery. The method used by Till *et al.* to develop a high-density TILLING population in rice is given in figure above. Seeds are treated with the mutagens and M_1 plants are allowed to self-pollinate and a single seed strategy is followed to M_2. M_2 plant is grown to provide DNA for mutation discovery and seed for banking. DNAs are pooled eightfold and arrayed in a two-dimensional format on 96-well plates. After PCR amplification of target genes, heteroduplexes are formed upon heating and annealing, and then digested using crude celery juice extract containing the CEL I nuclease. Cut strands are separated by denaturing polyacrylamide gel electrophoresis, and visualized by fluorescence detection using a Li-Cor DNA analyzer. The presence of cut products in two pools identifies the individual harboring the polymorphism. TILLING has worked well in a wide variety of plant species and until recently, has been one of the only available techniques for identifying *de novo* single nucleotide polymorphisms (SNPs) or point mutations in genomes for which the entire sequence is not available. In allotetraploid species such as peanut, TILLING is being used to improve selected characters such as the removal of allergens or improving quality traits. While the attempts to develop TILLING populations using mutagenized populations in allopolyploid annual species such as wheat, *B. napus,* soybean and peanut have been successful, researchers tend to use heteroduplex analysis by enzymatic cleavage to assess natural variation (EcoTILLING) for cross-pollinating annuals such as melon and sunflower and for forest tree species. Banana is the first perennial fruit species to be assessed by EcoTILLING.

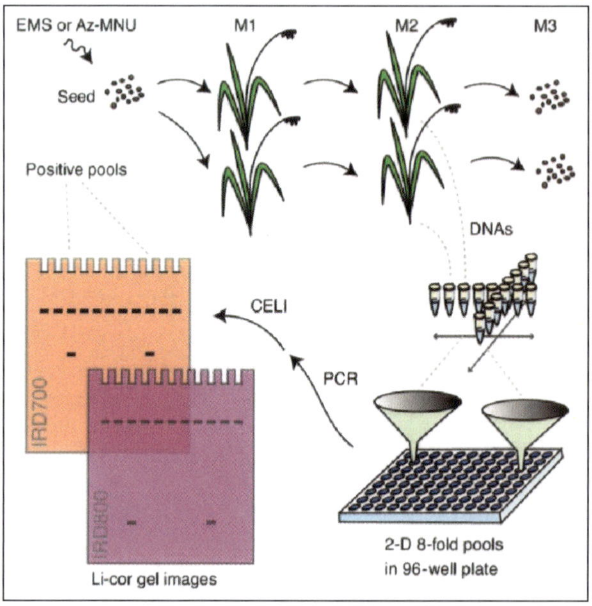

TILLING strategy for rice using endonuclease digestion.

One disadvantage of TILLING arises from relatively poor cleavage efficiency and exonuclease activity. Endonuclease digestion reactions and electrophoresis in classical TILLING are time-consuming and costly. Another disadvantage is its low sensitivity when screening genes with multiple small exons separated by larger introns as they rarely affect gene function. Also, a mutagenized population arrayed for TILLING is required, and for many species, establishment of such populations is time-consuming and expensive or can be complicated, for example when the desired cultivar is heterozygous and vegetatively propagated. To overcome these shortcomings, more robust technologies are being developed.

High-resolution melt analysis (HRM) distinguishes different PCR products by their thermal denaturation profiles. The melting temperature (T_m) of a PCR product depends on the GC content, length and sequence of the duplex. Small changes in the melting curve are indications of the heterogeneity in the sequence. Therefore, high resolution scanning and genotyping, enabled by new classes of saturation dyes coupled with high-throughput instrumentation, have contributed to the wider applicability of the technique, including plant mutation screening. HRM has been applied to cultivar identification of grape and olive by characterizing high-resolution melting of dinucleotide microsatellite repeats. Since then the method has been extended to capsicum, melon, tomato, wheat, potato and almond. Primers are being designed for amplification of sex markers in triploid hops. Li *et al.* developed a cost-effective dye (EvaGreen) for use in HRM analyses and successfully applied it for mutation screening of an EMS mutagenized maize population and mapped rice using SNP markers. The method has also been successfully applied for mutant screening in barley. Conserved regions of a gene often represent functional domains and have high sequence similarity between homoeologous loci, presenting difficulties in classical TILLING. Dong *et al.* demonstrated the usefulness of HRM as an alternative for screening mutations in conserved functional domains of homoeologous genes using an EMS mutagenized wheat TILLING population by screening mutations in the carboxyl terminal domain of the starch synthase II (*SSII*) gene. This method can also be used for SNP marker development and eco-TILLING in polyploid species.

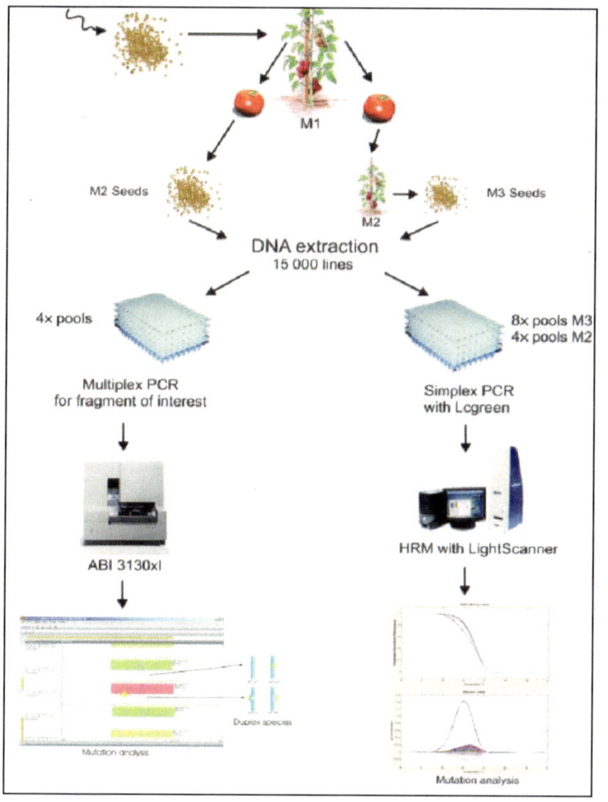

TILLING strategy for tomato using two non-enzymatic methods.

Another novel high-throughput, but low-cost mutant detection method based on capillary electrophoresis-single strand confirmation polymorphism (CE-SSCP) analysis was proposed by Julio *et al.* They applied it to a practical mutation breeding effort to produce nicotine-free tobacco, targeting two genes in the nornicotine biosynthetic pathway, *ADC1* and *NtabCYP82E4*. CE-SSCP was used to identify wild-type, heterozygous and homozygous plants in the M_2 progeny of EMS mutagenized tobacco for both genes. Homozygous plants for a truncated mutant of *NtabCYP82E4* displayed near-null nornicotine contents while the heterozygous mutant showed an intermediate level. Several other truncation and missense mutants of the same gene displayed very low-level nicotine-converter phenotype. The authors concluded that because the *NtabCYP82E4* nicotine *N*-demethylase mutated alleles are stably inactivated by point mutation, introgressed elite tobacco lines display a stable non-converter phenotype, whereas genetically modified tobacco through RNAi targeting may be silenced in some of the transgenic plants, thus reverting back to the wild-type. In tomato, Gady *et al.* used two non-enzymatic high-throughput screening techniques for TILLING; conformation sensitive capillary electrophoresis (CSCE) and HRM. This scheme simplifies TILLING as it avoids the time consuming endonuclease digestion reactions as well as gel electrophoresis but achieves an average mutation detection rate of 1.36 mutations per kb. For M_2 and M_3 populations, DNA was pooled 4 or 8 fold. For CSCE, after multiplex PCR amplification with fluorescent labeled primers, samples were directly pooled and loaded on capillaries filled with CAP polymer. Pools containing mutations were identified using Applied Maths' HDA peak analyser software. For HRM, pools were analysed for their product melting temperatures after PCR amplification in the presence of LC-Green+.

Development of Japanese pear mutants resistant to black spot disease using chronic gamma irradiation coupled with screening for colour change in leaf discs treated with AK toxin. (a) The disease susceptible leaves change to black and resistant material stay green when incubated with AK toxin; (b) diseased fruit; (c) fruit from a resistant cultivar. Head of Mutation Breeding Laboratory, Institute of Radiation Breeding (IRB), National Institute of Agrobiological Science, Japan.

With the continuous development of novel mutation detection techniques, targeted mutation breeding will become more practical in the future. In this respect, TILLING and

HRM allow sample pooling to reduce costs. A major constraint with polymerase chain reaction (PCR)-based techniques for mutation screening is their inability to selectively amplify low levels of mutations in a wild-type background. Li *et al.* demonstrated that replacing regular PCR with CO-amplification at Lower Denaturation temperature PCR (COLD-PCR) increases the mutation detection sensitivity up to 100-fold. By incorporating a synthetic reference sequence during COLD-PCR, Milbury *et al.* could enhance amplification of unknown mutant sequences even further. With next-generation sequencing (NGS) technology it is now conceivable to sequence candidate amplicons in multiple parallel reactions (SCAMPR), and the costs are comparable to TILLING.

Applications of Mutation Breeding and Impact of Mutant Cultivars

Crop Domestication and Adaptation to New Environments

Mutations have played an important role in plant domestication. Natural mutations and selection by man has been the driving force in the evolution of crop plants. Recent molecular studies show that duplications and polyploidization in particular are responsible for the evolution of many crop species. Crop yield increases attributed to changed plant architecture to suit modern agronomic practices such as dwarfing in cereals, bushy mutants of legumes, cotton, sunflower and castor are also the result of inducing and incorporating mutant genes into old cultivars. The reduction or complete removal of toxic substances such as alkaloids in lupins and glucosides in *Melilotus* species helped the introduction of these crops into agriculture.

Changing the product quality through mutation breeding also resulted either in new products from existing crops, such as the development of a new edible oil, linola, utilizing two induced mutants of flax isolated in Australia, or the overall improvement of product quality that resulted in the widespread use of a crop, such as in the case of a high oleic acid sunflower mutant that was bred in Russia and used extensively in the breeding programmes in North America and Europe to produce high oleic acid sunflower hybrids.

Biotic Stress Resistance

Through mutation breeding it has been possible to save some crops from diseases that had been devastating for growers. Powdery mildew resistance in barley was one of the first examples. At least 32 independent mutations in the *mlo* locus have been reported and, of these, most have been induced by chemical mutagens, five with radiation and one (*mlo*11) occurs naturally. Resistance to *Verticillium*-wilt in peppermint, a sterile allohexploid (*Mentha piperita*), was induced in 1959 by irradiation, and the resistance of newly bred cultivars was still effective after 30 years. Resistance to black spot disease (*Alternaria alternata* pv. Japanese pear), considered to be the most serious disease in Japanese pear (*Pyrus serotina* var.

culta), was induced in Japan in the 1960s through chronic irradiation of the cultivar 'Nijisseiki' and a resistant cultivar 'Gold-Nijisseiki' was released. The success of this programme led to radiation breeding of susceptible 'Shinsui' and 'OsaNijisseiki' cultivars resulting in resistant cultivars 'Kotobuki-Shinsui' and 'OsaGold'. The 'OsaGold' mutant has the added advantage of being self-compatible, eliminating the need to grow pollinators. The selection for disease resistance using AK toxin and a comparison of diseased and resistant Japanese pear cultivars is given in figure. The development and introduction of phytophthora-resistant sesame mutants through radiation breeding resulted in stabilizing the declining sesame yields in Sri Lanka.

There are many other examples of disease-resistant mutants developed either through direct selection after mutation induction or through cross-breeding with mutants. A good example of induced virus resistance is the cultivar 'NIAB Karishma', a cotton hybrid of the mutant cultivar 'NIAB 86', released in 1996 and resistant to cotton leaf curl virus. During the period 1997–2005, it brought additional income amounting to US$294.4 million to farmers in Pakistan. A number of pulse crop mutants with disease resistance have contributed to sustainable farming in India. Among these, bacterial pustule resistance in 'TAMS 38' soybean; powdery mildew resistance in mungbean mutants 'TARM 1', 'TARM 2', 'TARM 18' and 'TM 96-2'; yellow mosaic virus resistance in mungbean mutants 'TMB 37' and 'TJM 3'; and yellow mosaic virus resistance in urdbean mutant 'TU 94-2' are noteworthy. Combination of targeted mutation breeding techniques including the use of TILLING populations or HRC and the recently proposed strategy of disabling plant disease susceptibility genes (S-genes) will result in many more durably resistant cultivars developed through mutation induction.

Abiotic Stress Tolerance and Adaptability Traits

Mutants with tolerance to abiotic stresses such as salinity, acid soils, drought and toxic soil conditions have enabled breeders to develop crop cultivars adapted to adverse soil environments. Adaptability traits such as photoperiod and temperature response, changes in breeding systems of crops affecting cross-or self-incompatibility and male sterility are among many other characters that have been changed to produce new cultivars through mutation breeding. The gamma ray-induced 'Zhefu 802' rice cultivar with a shorter growing season, tolerance to cold and rice blast disease and high yield potential under low-input conditions was the most widely planted variety in China in 1986–1994. During this period its cumulative planted area reached 10.6 million ha. A concerted effort in mutation breeding of rice in Vietnam has resulted in a total area of 2.54 million ha of mutant varieties cultivated in the period 2000–2008 in the Mekong Delta Region, with eight rice mutant varieties adding an estimated return of US$374 million. These mutant varieties have early maturity allowing cultivation of three rice crops annually, lodging resistance, acid sulphate soil tolerance and resistance to major pests requiring two-to three-fold less pesticide sprays during the growing period. Similar successes with mutant varieties of rice have been reported from Australia, Thailand, Japan, India, Costa Rica and Myanmar.

As the *indica* type semi-dwarf donors did not have an acceptable grain quality, rice breeders in California successfully used induced mutations to produce semi-dwarf mutants in *Japonica* and *Javonica* backgrounds; 'Calrose 76' was the first direct release semi-dwarf cultivar produced through seed irradiation with gamma rays, released in 1976. Already in the early 1980s, the added value to rice farmers in California from the induced semi-dwarf mutant was US$20 million per year, as a result of a yield advantage of 14% over the tall *Japonica* types. Genetic evaluation showed that the semi-dwarfing gene in 'Calrose 76' was allelic to *sd1* in the indica 'green revolution' varieties derived from Dee Geo Woo Gen. Cross-breeding has resulted in 25 semi-dwarf varieties bred and released in the USA, Australia and Egypt that trace their ancestry to 'Calrose 76'. Mutation breeding has also contributed other semi-dwarf mutants non-allelic to *sd1*, waxy endosperm, early maturity and Basmati-type aromatic semi-dwarf mutants to the California breeding programme.

In cotton, a mutant cultivar 'NIAB 78' released for cultivation in Pakistan with high adaptability traits and better fibre quality recorded an additional cumulative income of US$612.4 million from 1983 to 2005. Although the economic analyses have not been published, sunflower became an intensive crop as a result of the many dwarf mutants developed in the early years of its cultivation. Radiation breeding is one way of adapting crops to new environments. When cross-breeding is not an option such as in the case of aromatic rice, mutation induction and screening in the new environment may enable quick adaptation of the crop.

Protoplast Fusion

Plant protoplasts can be fused and the fusion products cultured to produce somatic hybrid plants. This technique has been used to produce germplasm previously unavailable to the plant breeder. Several researchers have emphasized production of hybrids between distantly related, sexually incompatible species, but many of these hybrids are sterile, precluding incorporation into a breeding program. Hence, to transfer traits such as disease and herbicide resistance, emphasis has shifted to production of hybrids between more closely related species. Novel variation has been observed in such somatic hybrids due to segregation of mixed organelles, cytoplasmic and nuclear gene recombination, and somaclonal variation. The unique gene combinations made possible by protoplast fusion ensure that new plant varieties will soon be derived from somatic hybridization.

RNA Interference for Crop Improvement

RNA interference (RNAi) is a method of blocking gene function by inserting short sequences of ribonucleic acid (RNA) that match part of the target gene's sequence, thus no proteins are produced.

RNAi has the potential to become a powerful therapeutic approach toward targeted and personalized medicine. The range of diseases and disorders it might address is unprecedented; from cancer to cardiovascular diseases, neurodegenerative disorders and even HIV. Even more exciting is the potential of RNAi in agriculture. RNAi has provided a way to control pests and diseases, introduce novel plant traits and increase crop yield. Using RNAi, scientists have developed novel crops such as nicotine-free tobacco, non-allergenic peanuts, decaffeinated coffee, and nutrient fortified maize among many others.

Discovery of RNAi

Complete plant genome sequences has brought new dimensions in genetic modification to improve plant characteristics. Scientists believed that they could produce any gene products by just introducing foreign genes in plants, which was not always the case[1]. Plant biologists found out that introducing multiple copies of a gene that codes for purple petunia flowers led, not as expected to a deeper purple hue, but rather to plants with white or variegated flowers. Through an unknown mechanism, the introduced transgenes were silenced as well as the plant's 'purple-flower' gene. In another research, gene silencing was also observed when plants were infected with RNA viruses engineered to contain fragments of the plant gene.

The mechanism causing these effects was not known until American scientists Andrew Fire and Craig Mello discovered that injecting double stranded ribonucleic acids (dsRNA) into the worm *Caenorhabditis elegans* triggered the silencing of genes with sequences identical to that of the dsRNA. They called the phenomenon RNA interference. Re-examining the co-suppression pathway in petunia and virus-induced gene silencing revealed that all these processes led to the accumulation of dsRNAs, hence the RNAi pathway. Fire and Mello were awarded the 2006 Nobel price for Physiology or Medicine for their discovery.

In addition to its roles in regulating gene expression, RNAi is used as an immune response to infection and as a natural defense mechanism against molecular parasites such as jumping genes and viral genetic elements that affect genome stability. Specific types of bacteria have also been shown to trigger the RNAi pathway in plants.

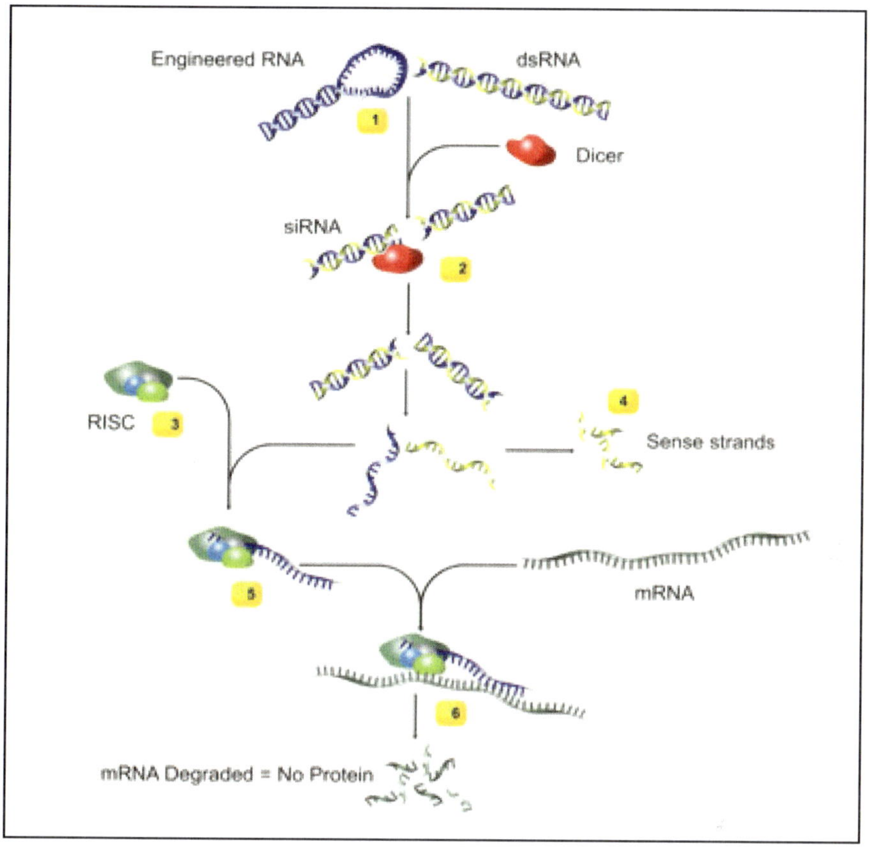

Mechanism of RNAi.

Working of RNAi

- The entry of long double stranded RNA, such as an introduced transgene, a rogue genetic element or a viral intruder, triggers the RNAi pathway of cells. This results in the recruitment of the enzyme Dicer.
- Dicer cleaves the dsRNA into short, 20-25 basepairs long, fragments, called small interfering RNA (siRNA).
- An RNA-induced silencing complex (RISC) then distinguishes between the two siRNA strands as either sense or antisense. The sense strands (with exactly the same sequence as the target gene) are degraded.
- The antisense strands on the other hand are incorporated to the RISC. These are used as guide to target messenger RNAs (mRNA) in a sequence-specific manner.
- Messenger RNAs (mRNA), which codes for amino acids, are cleaved by RISC. The activated RISC can repeatedly participate in mRNA degradation, inhibiting protein synthesis.

RNAi for Disease and Pathogen Resistance

Gene silencing was first used to develop plant varieties resistant to viruses. Engineered antiviral strategies in plants mimic natural RNA silencing mechanisms. This was first demonstrated when scientists developed Potato virus Y-resistant plants expressing RNA transcripts of a viral proteinase gene. Immunity has since been shown to other viruses such as the Cucumber and Tobacco Mosaic Virus, Tomato Spotted Wilt Virus, Bean Golden Mosaic Virus, Banana Bract Mosaic Virus, and Rice Tungro Bacilliform Virus among many others. In addition, plants can also be modified to produce dsRNAs that silence essential genes in insect pests and parasitic nematodes. This approach was used to develop root-knot nematode, corn rootworm and cotton bollworm resistant varieties.

RNAi for Male Sterility

RNAi has also been used to generate male sterility, which is valuable in the hybrid seed industry. Genes that are expressed solely in tissues involved in pollen production can be targeted through RNAi. For instance, scientists have developed male sterile tobacco lines by inhibiting the expression of TA29, a gene necessary for pollen development. RNAi was also used to disrupt the expression of Msh1 in tobacco and tomato resulting to rearrangements in the mitochondrial DNA associated with naturally occurring cytoplasmic male sterility.

RNAi and Plant Functional Genomics

A major challenge in the post-genomic era of plant biology is to determine the functions of all genes in the plant genome. Compared to other techniques, RNAi offers specificity and efficacy in silencing members of a gene or multiple gene family. In addition, the expression of dsRNAs with inducible promoters can control the extent and timing of gene silencing, such that essential genes are only silenced at chosen growth stages or plant organs.

There are several ways of activating the RNAi pathway in plants. The various RNAi techniques have advantages and disadvantages with respect to how persistent their effects are and the range of plants to which they can be applied. These include the use of hairpin RNA-expressing vectors, particle bombardment, Agrobacterium-mediated transformation and virus-induced gene silencing (VIGS).

Engineering Plant Metabolic Pathways through RNAi

RNAi has been used to modify plant metabolic pathways to enhance nutrient content and reduced toxin production. The technique takes advantage of the heritable and stable RNAi phenotypes in plants.

Table: Examples of novel plant traits engineered through RNAi.

Trait	Target Gene	Host	Application
Enhanced nutrient content	Lyc	Tomato	Increased concentration of lycopene (carotenoid antioxidant)
	DET1	Tomato	Higher flavonoid and b-carotene contents
	SBEII	Wheat, Sweet potato, Maize	Increased levels of amylose for glycemic management and digestive health
	FAD2	Canola, Peanut, Cotton	Increased oleic acid content
	SAD1	Cotton	Increased stearic acid content
	ZLKR/SDH	Maize	Lysine-fortified maize
Reduced alkaloid production	CaMXMT1	Coffee	Decaffeinated coffee
	COR	Opium poppy	Production of non-narcotic alkaloid, instead of morphine
	CYP82E4	Tobacco	Reduced levels of the carcinogen nornicotine in cured leaves
Heavy metal accumulation	ACR2	Arabidopsis	Arsenic hyperaccumulation for phytoremediation
Reduced polyphenol production	s-cadinene synthase gene	Cotton	Lower gossypol levels in cottonseeds, for safe consumption
Ethylene sensitivity	LeETR4	Tomato	Early ripening tomatoes
	ACC oxidase gene	Tomato	Longer shelf life because of slow ripening
Reduced allergenicity	Arah2	Peanut	Allergen-free peanuts
	Lolp1, Lolp2	Ryegrass	Hypo-allergenic ryegrass
Reduced production of lachrymatory factor synthase	lachrymatory factor synthase gene	Onion	"Tearless" onion

Prospects for RNAi

With RNAi, it would be possible to target multiple genes for silencing using a thoroughly-designed single transformation construct. Moreover, RNAi can also provide broad-spectrum resistance against pathogens with high degree of variability, like viruses. Recent studies have hinted possible roles of RNAi-related processes in plant stress adaptation.

Although much progress has been made on the field of RNAi over the past few years, the full potential of RNAi for crop improvement remains to be realized. The complexities of RNAi pathway, the molecular machineries, and how it relates to plant development are still to be elucidated.

Transgenics

Gene is the fundamental unit of heredity in any organism. Mendel first proposed this in 1865 but he called it 'factor'. Later, in 1909 Johansson renamed 'factor' as gene. Gene is an active segment of DNA. The isolation and manipulation of an organism's genome for the betterment of mankind is termed genetic engineering, a fast growing science in the field of biotechnology. One among the products of genetic engineering is transgenics.

Simply, *transgenics* is *transferring foreign genes* from one organism to another organism of interest. The transgenic plants and animals derived thus are frequently referred to as genetically modified organisms (GMOs). The first transgenic plant was tobacco- developed in 1983. There are more than 50 other plant species where foreign genes have been transferred viz., tomato, potato, sunflower, cotton, carrot, grapes, etc. Initially, the production of trangenics was restricted to dicotyledonous plants for the reasons detailed below. But, now it has been extended to several monocots like wheat, maize, rice, oats, etc.

The need of Transgenics

Classical genetics, as applied to plant breeding, cannot be separated from other improvements in agricultural production.

It has been estimated that genetic improvement alone accounts for 50% of the increased harvest.

Conventional breeding techniques aim to introduce genetic diversity into desired plants so as to evolve superior plants. This involves many time-tested procedures. However, evolving elite varieties takes comparatively a very long time, even up to 8 years. Further, in conventional breeding, the transfer of desirable genes from unrelated or wild relatives is limited for any of the reasons below:

- Failure of pollen germination;
- Reduced seed set due to slow pollen tube growth;
- Sexual incompatibility;
- Abnormality of seed development;
- Hybrid inviability (seedling lethality);
- Hybrid sterility; and
- Transfer of undesirable characters from the wild species.

Genetic engineering serves as a promising alternative tool for overcoming the above limitations. The addition of molecular biology and genetic engineering techniques to conventional methods allows the plant breeders to accomplish objectives that are impossible through conventional plant breeding. The tools of biotechnology are unlikely to replace any of the time-tested methods of crop improvement. Rather, biotechnology aims to increase the efficiency of various methods of conventional breeding.

Development of transgenic plants involves the following three steps:

Construction of Recombinant DNA

The first step involves the purification of the desired genes (DNA sequence) and construction of a recombinant DNA. First the DNA from the donor organism is located and isolated.

Isolation of genes is generally done by probing a library for genes of interest (GOI). Other methods used are T-DNA tagging, transposon tagging, positional cloning, etc. Passenger DNA is sometimes chemically synthesized from nucleotides deduced by the amino acid sequence of proteins. Normally, GOI thus identified cannot be directly transfered into the plant tissues but needs a vehicle.

The GOI have to be cloned with any one of the vectors like plasmids, bacteriophages, yeast artificial chromosome (YAC), bacterial artificial chromosome (BAC), etc. with introducing certain specific promoter and terminator sequences. Thus the recombinant DNA is constructed. All these steps involve splicing uand joining of DNA molecules, which are made possible by restriction endonucleases and ligases, respectively.

Transformation of Gene of Interest into Plant

Transformation of the gene of interest into a plant may be done using any one of the following methods.

Agrobacterium-mediated transformation: Agrobacterium tumefaciens is a soil bacterium and was the first to be used to transform plant tissues. The bacterium causes tumour in wounded gymnosperms and dicot angiosperms as they possess a single copy of large tumour inducing (Ti) plasmid in the cell. The Ti plasmid also contains transfer DNA or T-DNA, which transfers itself to the host and causes tumour. These tumourcausing genes are removed and our gene of interest is inserted into the T-DNA. When *A. tumefaciens* with the recombinant DNA is made to infect plant cells, the DNA gets transformed into the plant genome and expresses itself later.

Direct gene transfer: Direct gene transfer is the direct delivery of the recombinant DNA into plant protoplasts or calli either physically or chemically.

In the chemical method, the recombinant DNAs are transferred into plant protoplasts under the presence of calcium phosphate or polyethylene glycol (PEG). In the physical method called electroporation, electrical impulses of high field strengths are used to reversibly permeabilize cell membranes to facilitate the uptake of large molecules, including foreign DNA.

Another technique that has attained major importance is the biolistic approach (particle gun bombardment) developed by Sanford and others. In this approach, microscopic tungsten or gold particles (4 mm) carrying the recombinant DNA are accelerated to 400 ms^{-1} and allowed to penetrate intact cell walls of calli or protoplasts.

Reproduction and Regeneration of Transgenic Plant Cells

The transformed cells are selected and cultured on an artificial medium to grow into callus. Under the appropriate ratio of auxin and cytokinin, several plant species can be regenerated into whole plants from the calli, the phenomenon called totipotency.

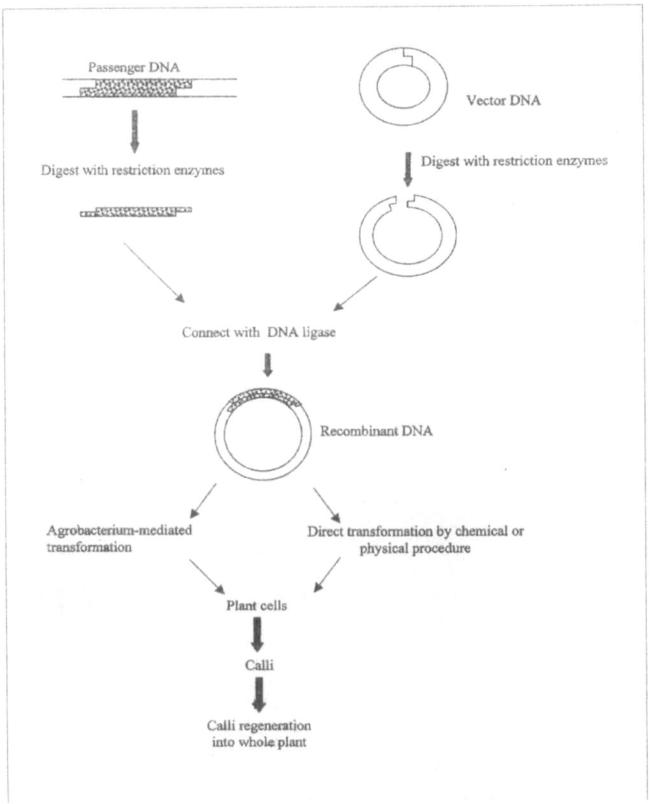

Schematic representation of transformation of gene into plant.

Utilization of Transgenic Plants in Agriculture

Many economically important genes were engineered into crop plants worldwide. But a major contribution has been made in the field of crop protection.

Insect-resistant Transgenic Plants Expressing Bt Toxin

Bacillus thuringiensis (universally abbreviated as Bt) is a gram positive soil bacterium. The bacterium produces crystalline toxins on sporulation which are coded by cry genes. When consumed by the insect larvae the toxins affect the midgut epithelium resulting in death of the larvae. Different strains of Bt produce toxins effective against different insect species. Lepidopteran larvae are the most susceptible followed by Coleoptera and Diptera.

To produce an insect-resistant transgenic plant, the DNA sequence coding for the toxin is to be removed from the bacterium. With suitable promoter and terminator sequences recombinant DNA is constructed. This gene construct is introduced into plants using Agrobacterium tumefaciens as vector. Transgenic tobacco, potato and tomato plants resistant to various pests have thus been transformed.

Herbicide-resistant Transgenic Plants

Broad spectrum and non-selective herbicides are used to kill unwanted plants or weeds in agriculture. But these are equally capable of killing crop plants. Hence, crop plants resistant to broad-spectrum herbicides are to be developed so that we can spray herbicides on all weeds growing between crops. Transgenic plants resistant to many herbicides have been developed of which glyphosate is important.

Glyphosate is a non-selective herbicide. It is a specific inhibitor of 5-enol pyruvyl shikimic acid-3-phosphate synthase (EPSPS) which catalyses one of the steps in the shikimic acid pathway. This is the key pathway, present only in plants and microorganisms, leading to the synthesis of many amino acids playing a vital role in protecting the plants against pest and diseases. As a result of inhibition by the herbicide massive accumulation of shikimic acid and the lethal depletion of end product aromatic amino acids occur.

To evolve crops resistant to glyphosate, insertion of EPSPSinsensitive genes was considered. Mutagenesis of yeast, Saccharomyces typhirnurium, yielded a glyphosate-resistant mutant possessing an insensitive EPSPS encoded by aroA gene. A gene construct with aroA gene and octopine or monopine synthase promoters was transferred into tobacco and tomato mediated by Agrobacterium. Transformed plants showed significant tolerance to glyphosate.

Transgenics for Improved Quality

The quality of agricultural produce has been improved utilizing the transgenic technique in many instances. Some examples are:

The shelf life and solid content of tomato fruits has been altered by inhibiting ethylene production or by decreasing polygalacturonase activity by antisense RNA technique.

Starch content of potato tubers has been increased by heterologous expression of mutated ADP-glucose pyrophosphorylase gene from E. coli.

The expression of synthetic genes encoding proteins rich in essential amino acids like lysine and methionine, along with normal protein production within the potato tuber, may increase the overall nutritional quality of the potato.

In chrysanthemum, flower colour was improved using cloned pigment-biosynthesis genes.

Limitations of Transgenics

Transgenic technology can be very attractive and has many advantages compared with other conventional procedures. But it has certain drawbacks too. Some of these are listed below.

Technical Limitations

- Transgenic expression is generally unpredictable and varies depending on the transgene copy number and site of integration.
- Transgene inactivation and silencing by the host system has emerged as a potential problem in the use of heterologous system.
- The introduced transgene sometimes suppresses endogenous gene expression by a phenomenon called cosuppression.
- Expression of the transgene may turn cytotoxic and inhibitory to host growth and metabolism.
- Somaclonal variation occurs to a certain degree in all trans-genes due to the stress induced by tissue culture and transformation procedures.
- There is a lack of technique for transferring multiple genes by transformation.
- The phenomena of recalcitrance in certain plant species limits transformation, transgene integration and regeneration of the plants.

Social Limitations

- Toxicity of the marker gene/gene product.
- Vector-mediated horizontal gene transfer and recombination to create new pathogenic bacteria.
- The spread of transgenic crops by pollen transfer to sexually compatible species may contaminate the native genetic information.
- Genetically modified plants with marker genes do limit non-target insect life that comes in contact with plant exudates, such as sap or pollen.

- There are chances for transfer of genes from herbicideresistant crops to wild or semi-domesticated relatives which would create super weeds.
- Insect pests could quickly develop resistance to transgenic crops.
- There are possibilities for vector recombination to generate new virulent strains of virus, especially in transgenic plants engineered for viral resistance with viral genes.
- Transgenic plants with foreign genes would surely affect various trophic level interactions through the food chain.

Genome Editing

The precise modification of pre-selected sequences in the plant genome is the holy grail of plant breeding because it allows the rapid introduction of genetic diversity and accelerates the generation of improved varieties, especially in polyploid crops, which otherwise need to undergo lengthy programs of crossing and screening. The development and application of genome-editing tools in recent years has, therefore, revolutionized basic research in plant biology and the generation of new plant varieties. Although the molecular basis of genome editing in plants has been comprehensively discussed in the literature, the practical aspects have received comparatively little attention. Accordingly, in this part, we emphasize some of the technical and practical aspects of genome-editing technologies for crop improvement.

Genome-editing Tools

Programmable Sequence-specific Nucleases

There are three major genome-editing techniques categorized by mechanism of action, and by far the most commonly used in plants is the targeted generation of DNA double strand breaks (DSBs) using programmable sequence-specific nucleases followed by DNA repair via one of two major endogenous pathways. The error-prone non-homologous end joining (NHEJ) pathway joins DNA ends and is often accompanied by the insertion or deletion (indel) of short stretches of nucleotides at the junction. When the DSB is within a coding sequence, the resulting mutation often causes a loss of function, which can be exploited to determine gene functions (reverse genetics) or to abolish undesirable traits to improve crops. In contrast, homology-directed repair (HDR) is a more accurate pathway but it requires the presence of a DNA template with homology to the sequences upstream and downstream of the DSB. By providing a donor DNA template with homology to the target site, this pathway can be exploited for the precision engineering of endogenous genes or for the addition of genes or other sequences at predetermined genomic loci.

The targeted induction of DSBs is achieved using programmable nucleases. The most common nucleases for genome editing are zinc finger nucleases (ZFNs), transcription activator-like effector nucleases (TALENs), and Cas9, the latter being part of the clustered regularly interspaced short palindromic repeat (CRISPR)/Cas9 system. The three classes of nucleases differ in structure, activity, and enzymatic mechanism, resulting in differences in target selection, efficiency, specificity, and mutation signature. We will focus on these practical aspects of each nuclease class and refer readers to other comprehensive reviews for information on the history of each system and the development of improved variants.

ZFNs and TALENs are artificial enzymes consisting of a series of DNA-binding domains (zinc fingers and TAL effector domains, respectively) fused to the sequence-independent catalytic domain of the type IIS restriction endonuclease FokI. In its natural form, FokI must dimerize to cleave DNA, so both ZFNs and TALENs also function as dimers and, like FokI, generate DSBs with 5j cohesive overhangs.

One zinc finger module binds a triplet of nucleotides. A single zinc finger would, therefore, lack the specificity to bind a unique genomic target, but typical constructs contain three or four fingers, which translates to a ZFN target sequence of 18–24 bp (9–12 bp per half-site). This length is sufficient to target unique sites even in the large genomes of higher eukaryotes. The target site can be any length in principle, but the context-dependent assembly of ZFN modules limits the size in practice. Unlike zinc fingers, each TAL effector recognizes a single nucleotide. Engineered TAL effector domains are typically designed to recognize 15–30 nucleotides, making a total of 30–60 nucleotides for one TALEN pair. Therefore, TALENs are generally considered to bind with greater specificity than ZFNs, even though larger TAL effector domains are more likely to tolerate mismatches. To further minimize off-target events and associated cellular toxicity, the FokI nuclease dimerization interface has been engineered to force heterodimer formation, and this variant is routinely used in ZFNs and TALENs.

In terms of target choice, publicly available ZFN module libraries can be used to prepare functional ZFN pairs that match one site in every ~100 bp of random genomic DNA. This means that target sites are available for the disruption of most conventional genes, but it may be more difficult to target small non-coding RNA genes, short regulatory regions, or genomic sites that happen to fall between the available targets. In contrast, the selection of target sites for TALENs is restricted only by the requirement for a thymidine at the first position, so TALEN targets can be found on average every 35 bp.

The half-sites recognized by ZFNs are typically separated by 6–8 bp depending on the design of the fusion protein. For TALENs, the spacer length is usually 15–30 bp. In the absence of a repair template, both nucleases predominantly generate indels a few base pairs in length, but ZFNs tend to produce a larger proportion of insertions than deletions, probably because the short overhangs resulting from the shorter spacers can be more efficiently filled in before ligation. The generation of more insertions than deletions is a potential drawback in terms of regulation because the added DNA might be considered a novel sequence, but this aspect has yet to be explored in practice.

The number of publications reporting the use of ZFNs and TALENs in plants is rather limited and tends to favor TALENs, but the editing efficiency that can be achieved with both types of nuclease (i.e., the number of correctly edited events as a proportion of all attempts) appears to be similar. Under appropriate experimental conditions, high editing efficiencies are possible with both ZFNs and TALENs. In maize (*Zea mays*), for example, the reported efficiency of gene targeting with ZFNs was 40% compared to 40–60% with TALENs. Lower efficiencies reported in other cases can probably be attributed to the experimental conditions and/or the choice of target sequence. TALENs are significantly larger and more repetitive than ZFNs, which can affect the efficiency of intracellular nuclease delivery and in turn the overall efficiency of editing. The requirement for larger constructs can also exclude the use of viral vectors, which have a limited insert capacity.

Where specifically investigated, ZFNs did not induce off-target mutations in plants and TALENs only rarely caused cytotoxicity due to off-target cleavage. However, the authors in each case only looked for mutations at predicted off-target sites, whereas genome-wide screening is required for unbiased analysis. The identification of potential off-target sites for ZFNs and TALENs is challenging (particularly for ZFNs) because sequence specificity is conferred by protein–DNA interactions that are often context-dependent and difficult to predict. A final drawback of ZFNs and TALENs is the requirement of different dimeric proteins specific for each target site, which limits their practical use for multiplexing (the simultaneous introduction of DSB at multiple sites) unless the target sites are closely related.

The most recent addition to the toolbox of programmable nucleases (and the most widely used in plants) is Cas9 from *Streptococcus pyogenes* (SpCas9), which is part of the CRISPR/Cas9 system. Cas9 is a monomeric nuclease that forms a ribonucleoprotein (RNP) complex with a chimeric guide RNA (gRNA). The latter confers sequence specificity by providing a 20-nucleotide sequence complementary to the target site (known as the protospacer). The Cas9 enzyme possesses two nuclease domains, each cleaving one strand of the target sequence three nucleotides upstream of the protospacer adjacent motif (PAM) to generate blunt ends. The only constraint for the design of the gRNA is the PAM, which needs to be present at the 3j end of the target sequence. For SpCas9, this sequence is defined as 5j-NGG-3j. The in silico analysis of plant nuclear genome sequences (including monocots and dicots) has identified 5–12 NGG-PAMs on average for every 100 bp. Given the relatively short target sequence and the high frequency of PAM sites, it can be challenging to identify specific targets, especially in crops with large and highly repetitive genomes such as maize. Cas9 variants from other species such as *S. aureus* (SaCas9, recognizing the less frequent NNGRRT-PAM and SpCas9 mutants that recognize non-canonical PAMs, can broaden the range of CRISPR/Cas9 genome-editing targets in plants with complex genomes.

One major advantage of CRISPR/Cas9 over ZFNs and TALENs is that the Cas9 enzyme does not need to be engineered at the protein level to recognize different targets. Target specificity is conferred entirely by the spacer region of the gRNA, and the sequence can be modified using standard molecular biology methods. Given the simplicity of the CRISPR/Cas9 system, it has superseded ZFNs and TALENs in research applications and a large body of literature has accumulated, describing the use of this system in many different plant species. Furthermore, by providing multiple gRNAs simultaneously, the CRISPR/Cas9 system can be used to target different genes in parallel, including unrelated genes.

The CRISPR/Cas9 system is at least as efficient as ZFNs and TALENs in cereal crops and for most species, there are examples of editing efficiency approaching 100%. In general, targeting one gene at two positions increases the overall mutation frequency and allows the recovery of homozygous mutants in one generation. The ease of multiplexing with the CRISPR/Cas9 system is therefore an advantage for the generation of knockouts using this dual-gRNA approach. In contrast to ZFNs and TALENs, Cas9 generates blunt DSBs that are typically repaired by the formation of small (usually 1-bp) indels, leading to the frequent recovery of frameshift mutants when the target site is within an exon.

More recently, the CRISPR/Cas12a (originally named Cpf1) system was discovered and also developed into a genome-editing tool. Cas12a recognizes a TTTN-PAM preceding the target sequence, which facilitates the targeting of AT-rich regions and reduces the likelihood of off-target mutations in GC-rich genomes, adding more flexibility to the application of CRISPR/Cas systems in plants. Cas12a produces a staggered cut with 5j overhangs of five nucleotides more distal to the PAM, which in turn favors the generation of deletions 6–13 bp in length, considerably larger than the mutations generated by Cas9. These features further broaden the sequence space that can be targeted using CRISPR/Cas systems. A recent study in rice suggests that several Cas12a variants are temperature sensitive and that editing efficiencies can be increased substantially when plants are grown at 28 ◦C instead of 22 °C.

The frequency of off-target mutations generated by CRISPR/Cas systems has been raised as a concern, even though the specificity of CRISPR/Cas9 in plants appears to be higher than in mammals. Unexpected DSBs have been reported for only a minority of gRNAs, even when whole-genome sequencing has been used to screen for off-target mutations. Several high-fidelity variants of SpCas9 and SaCas9 have also been developed, along with strategies to increase the length of the recognition sequence by engineering an inactive Cas9 enzyme fused to the FokI nuclease domain, similar in principle to ZFNs and TALENs. In most cases, the likelihood of off-target mutations can be reduced by optimizing the experimental setup (discussed in more detail below) and by the careful design of the gRNA. Recently, a further major improvement in the prediction of off-target sites in plants has been achieved by developing better alignment algorithms and taking into account rare but important off-target sequences.

Base Editors

Given the unpredictable outcome of DSB repair via NHEJ and the low efficiency of HDR, researchers have sought new methods to introduce point mutations without the need for DSBs or donor templates, resulting in the development of base editors as a new tool for genome editing. To achieve the conversion of cytidine to uridine, a cytidine deaminase was fused via a linker to the N-terminus of a semi-active version of Cas9, in which one of the nuclease domains is inactivated by the mutation D10A to generate a nickase (nCas9). The combination of the nickase and cytidine deaminase triggers DNA mismatch repair, resulting in the targeted conversion of a C:G base pair to T:A. In addition to base substitutions, indels can sometimes occur as a consequence of the DNA nicks induced by nCas9 on the non-edited strand. The first generation of base editors had an editing window of a few base pairs, causing unwanted bystander mutations. By controlling the length and flexibility of the linker, a high-precision base editor has recently been reported that can selectively edit a single C at a specific position with high accuracy and efficiency. Recently, base editors based on Cas9 variants with high fidelity and relaxed PAM requirements have been developed.

The conversion of A:T to G:C is more challenging because there are no known adenine deaminases that act on DNA, but this has been addressed by the directed evolution and engineering of tRNA adenosine deaminases to accommodate DNA substrates. A tRNA adenosine deaminase has recently been adapted for base editing in plants by directed evolution, achieving A:T to G:C conversion at frequencies of up to 7.5% in protoplasts and 59.1% in regenerated rice (*Oryza sativa*) and wheat (*Triticum aestivum*) plants. This base editor has a deamination window extending from positions 4 to 8 of the protospacer. In mouse embryos, unbiased mutation screening has recently shown that base editing causes off-target mutations at a 20-fold higher frequency than Cas9, raising concerns over the use of this technology for sensitive applications. No significant conversion has been observed at predicted off-target sites in plant DNA, but genome-wide screening using unbiased methods is needed to properly assess the specificity of base editors. A recent study reported low but detectable off-target tRNA adenosine deaminase activity against cellular RNAs in mammalian cells and described base editor enzymes with reduced DNA and RNA off-target activity. Even so, base editing is more efficient for the introduction of point mutations than alternative approaches such as TILLING or gene targeting by homologous recombination. Multiplexed base editing without any loss of efficiency has been reported in mammals but not yet in plants.

Oligonucleotide-directed Mutagenesis

Mutagenic DNA oligonucleotides 20–200 nucleotides in length have been delivered into plant cells to introduce point mutations in target genes, an approach known as oligonucleotide-directed mutagenesis (ODM). This technique harnesses the endogenous HDR pathway to correct mismatches generated by pairing the exogenous oligonucleotide, which carries the desired sequence, to its near-complementary target site in the genome. The oligonucleotide therefore acts as both a mutagen and a DNA repair template. Mutagenesis can be achieved using standard single-stranded DNA oligonucleotides (ssODNs) but these have a short intracellular half-life, and their efficiency has therefore been improved by stabilizing modifications. The modified variants include chimeraplasts (duplexes of DNA and methylation-modified RNA), phosphorothioate-modified ssODNs, and ssODNs with a 5j Cy3 label and a 3'idC reverse base modification.

The efficiency of ODM is generally rather low and positively correlates with oligonucleotide length, at least in the case of ssODNs, where increasing the length to 200 nucleotides achieved precise editing frequencies of up to 0.05% at a transgenic locus in Arabidopsis (*Arabidopsis thaliana*) protoplasts. Chimeraplasts did not increase mutation frequencies above the level of spontaneous mutations in tobacco (*Nicotiana tabacum*) or rapeseed (*Brassica napus*). However, targeted mutation frequencies could be enhanced by ODM in concert with nonspecific DSB-inducing reagents such as antibiotics, or sequence-specific nucleases such as TALENs and CRISPR/Cas9 in Arabidopsis and flax (*Linum usitatissimum*). Off-target mutations could be generated by ODM due to oligonucleotide recombination or off-target mismatch repair, but such events have not been reported. ODM theoretically allows multiplexing, i.e., multiple conversions at several targets within a single gene or the simultaneous conversion of multiple targets in a single cell.

Outcomes of Genome Editing

There are five classes of genome modifications that can be induced using the tools described above, although not all outcomes are possible with all tools. All five types of modification have been reported in crop plants, and each is discussed below with some examples of gene modifications associated with relevant traits and outcomes. Figure presents an overview of the different genome-editing tools, the potential outcomes of genome editing in each case, and examples of modified crop traits.

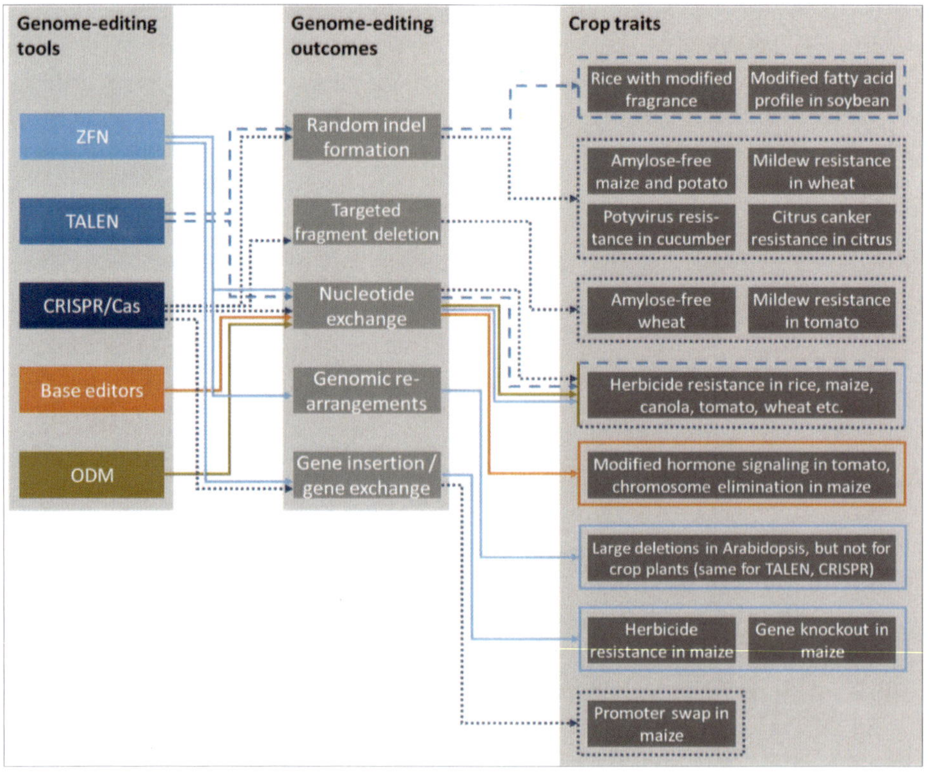

Overview of genome-editing tools, the possible genetic outcomes in each case, and examples of crop traits generated using these tools. The colored arrows and boxes link published crop trait examples with the associated genome-editing tool and outcome.

Random Indel Formation

When a single DSB is induced in higher plants, the most frequent outcome is the formation of indels around the cleavage site. When the indels introduced by erroneous break repair lead to a frameshift—which theoretically is the case for 2/3 mutations within the open reading frame of a gene—the resulting transcript usually contains premature stop codons. These are recognized by endogenous quality control processes and can initiate nonsense-mediated mRNA decay and thus result in the functional knockout of the mutated gene. This is the simplest form of genome editing. Most applications of sequence-specific nucleases (particularly the CRISPR/Cas9 system) have concerned the generation of indels to achieve gene knockout. Some examples are discussed.

TALENs have been used to knock out the rice *OsBADH2* gene encoding a betaine aldehyde dehydrogenase. Homozygous plants of the non-fragrant rice variety Nipponbare with mutations in this gene emitted a desirable fragrance due to the accumulation of 2-acetyl-1-pyrroline, similar to natural fragrant varieties such as Indian Basmati and Thai Jasmine, which usually trade at higher market prices. TALENs have also been used in soybean (*Glycine max*), a staple crop cultivated for the high protein and oil content of its seeds, to modify the fatty acid profile. Soybean oil naturally contains a high percentage of the polyunsaturated fatty acids linoleic acid (18:2) and linolenic acid (18:3) making it sensitive to oxidation. Therefore, soybean oil has been partially hydrogenated to reduce the quantity of polyunsaturated fatty acids, but this process generates trans-fatty acids with negative health effects. Such trans-fatty acids resulting from the partial hydrogenation of food ingredients have therefore been banned by the FDA. To reduce the levels of linoleic and linolenic acid, the genes encoding fatty acid desaturases FAD2 and FAD3 were disrupted using TALENs. Homozygous soybean plants carrying the knockout alleles *fad2-1a*, *fad2-1b*, and *fad3a* produced oil with <3% linoleic acid and linolenic acid (compared to 51% and 8%, respectively, in wild-type plants) but much higher (> 80%) levels of oleic acid (18:1) than wild-type plants (23%). Due to the lower levels of polyunsaturated fatty acids, the seed oil from these events is less sensitive to oxidation and does not require processing by partial hydrogenation.

Starch is the main carbohydrate storage molecule in plants and has many applications in addition to its use in food and beverages. Naturally, starch occurs in two different forms (amylose and amylopectin), but for certain industrial applications, it is desirable to have a homogenous starch composition rather than a mixture. Therefore, efforts have been made to engineer maize and potato (*Solanum tuberosum*) plants that are devoid of amylose and accumulate only amylopectin, a phenocopy of the natural maize mutant *waxy*. Granule-bound starch synthase I (GBSSI) is the key enzyme required for amylose synthesis, and has been targeted in tetraploid potato plants by transfecting protoplasts with preassembled Cas9/gRNA RNPs. Using this approach, 2.3% of all regenerated shoots contained mutations in all four alleles of the *GBSSI* gene, highlighting the power of sequence-specific nucleases in polyploid crops.

Engineering pathogen resistance is another major goal of crop improvement. Sequence-specific nucleases have been used successfully in this context, including the generation of wheat plants resistant to *Blumeria graminis* f. sp. *tritici*, the fungal pathogen responsible for powdery mildew disease. Wheat plants with a non-functional mildew resistance locus (*MLO*) are naturally resistant to the pathogen, but the locus has been difficult to target using traditional mutagenesis techniques because bread wheat is an allohexaploid species with six *MLO* homeoalleles at three loci (*TaMLO-A1*, *TaMlo-B1*, and *TaMlo-D1*). Given the ability of single TALEN pairs and single gRNAs to target multiple conserved sequences simultaneously, both methods have been used to generate knockouts of all six homeoalleles (*tamlo-aabbdd*) resulting in mildew-resistant wheat varieties. Similarly, CRISPR/Cas9 technology has been used in cucumber (*Cucumis sativus*) to knock out the gene encoding translational initiation factor eIF4E, which is essential for the Potyvirus infection cycle, generating plants with broad virus resistance. Finally, disease resistance has also been achieved by using CRISPR/Cas9 in Wanjincheng orange (*Citrus sinensis* Osbeck) plants to target the promoter of the susceptibility gene *CsLOB1*, resulting in plants with enhanced resistance to citrus canker.

Targeted Fragment Deletion

When two or more DSBs are induced within the same gene, it is possible to achieve gene knockout via the targeted deletion of the intervening sequence. Although theoretically possible with any sequence-specific nuclease, this form of genome editing is most easily achieved with the CRISPR/Cas9 system, which is highly amenable to multiplexing. There are two main advantages associated with this approach: First, when deleting a larger fragment of DNA, the chances of disrupting gene function are higher; and second, analysis of the events can be greatly simplified when the difference in size between wild-type and mutated PCR amplicons can be assessed simply by gel electrophoresis or melt analysis, rather than fragment digestion or sequencing. The drawback is that the frequency of fragment deletion depends on the efficiency of the nucleases, and often results in indels at one or both cleavage sites rather than removal of the intervening sequence. Increasing the number of gRNAs can also increase the number of potential off-target sites.

In practical examples of this approach, scientists at Pioneer Hi-Bred used CRISPR/Cas9 to inactivate the enzyme GBSSI in elite maize germplasm by deleting a large portion of the *wx1* locus using two gRNAs. Disease resistance has also been achieved using this approach, for example in the tomato (*Solanum lycopersicum*) *Mlo1* locus, which was targeted using two gRNAs to generate a 48-bp deletion. Homozygous and biallelic deletion mutants were resistant to the powdery mildew fungus *Oidium neolycopersici*.

Targeted Nucleotide Exchange

Precise nucleotide exchanges (often described as gene conversion or allele replacement events if they are driven by homologous recombination) can be used not only to achieve gene knockout (e.g., by introducing an early termination codon), but also to generate gain-of-function mutations or regulatory mutations that modulate gene expression.

All three of the genome-editing tools described above can be used to achieve nucleotide exchanges, but unassisted sequence-specific nucleases generate such mutations only rarely, and the replacement is random. Therefore, targeted nucleotide exchange using sequence-specific nucleases is generally achieved by providing a repair template of either single-stranded or double stranded DNA carrying the desired allele, to promote DSB repair via the endogenous HDR pathway (i.e., gene conversion). Higher plants overwhelmingly favor NHEJ over HDR, so the frequency of desired point mutations is typically rather low. The few examples that have been reported mainly concern the generation of a selectable trait such as herbicide resistance to favor the isolation of precise editing events. TALENs and the CRISPR/Cas9 system have been used for nucleotide exchange in the rice gene encoding acetolactate synthase (*ALS*), generating plants resistant to bispyribac-sodium. Furthermore, ZFNs have been used for nucleotide exchange in the wheat *ALS* gene to confer resistance to imidazolinone herbicides By delivering the CRISPR/Cas9 components either as DNA or RNP in combination with single-strand oligonucleotides, the maize *ALS2* gene was also edited and chlorsulfuron-resistant plants were obtained.

ODM can also be used to promote nucleotide exchange. Using this technology, Cibus has already produced a commercial sulfonylurea-tolerant canola line (SU Canola) and several other similarly modified crops are in the pipeline: Rice, flax, potato, wheat, maize, cassava (*Manihot esculenta*), and peanut (*Arachis hypogaea*). The ODM approach can also be combined with sequence-specific nucleases so that the ssODN acts as the repair template for the DSB. This strategy has been used to introduce nucleotide exchange in the flax *EPSPS* gene encoding 5j-enolpyruvylshikimate-3-phosphate synthase, conferring glyphosate resistance. Notably, the editing efficiency was sufficient to allow the regeneration of edited plants without selection.

Another way to achieve targeted nucleotide exchange is the use of base editors. For example, base editors incorporating cytidine deaminase have been used to generate rice varieties with precise exchanges in genes conferring herbicide resistance and senescence-related traits, and tomato plants with the anticipated modification in a gene controlling hormone signaling. In maize, the same approach has been used to introduce single-nucleotide exchanges in the CENP-A targeting domain of CENH3, a protein required for uniparental chromosome elimination during the production of double haploids. More recently, base editors incorporating adenosine deaminase have been used to modify the pathogen-responsive phosphorylation site in several endogenous rice genes, to generate rapeseed plants with a delayed flowering phenotype, and to confer herbicide resistance in rice.

Genomic Rearrangements

The introduction of two DSBs in the same gene can lead to targeted DNA fragment deletion as discussed above. However, if the two DSBs are further apart on the same chromosome, the outcome can be a major cytogenetic deletion or inversion, and if the DSBs are on different chromosomes, then the resolution can generate a translocation event. Although rare, such genomic rearrangements offer interesting opportunities to breeders: Large deletions could be used to remove entire gene clusters, translocations can create new linkages between interesting traits or break undesirable linkages, and inversions on homologous chromosomes could help to transfer traits from wild relatives to elite cultivars.

Large chromosomal deletions (up to ~9 Mb) and inversions (~10.6 kb) have been generated in Arabidopsis using ZFNs. In rice, large chromosomal deletions of up to 245 kb were generated in protoplasts and regenerated plants using CRISPR/Cas9, but their transmission to the T1 generation was not investigated. Furthermore, a recent preprint reports the CRISPR/Cas9-mediated duplication and meiotic transmission of a 2.3-kb DNA fragment in Arabidopsis. In rice, the inversion of a 1.3-kb DNA fragment was achieved with a frequency of 1% using TALENs, and the inversion of a ~300-bp fragment between two DSB generated by CRISPR/Cas9 was observed in one of nine tested gRNA pairs. Although theoretically possible, heritable larger inversions and translocations have not yet been reported in crop species.

Gene Insertion and Gene Exchange

The DSBs generated by sequence-specific nucleases can also be exploited to insert larger sequences, including complete genes, either by direct fusion to the free ends (NHEJ) or by recombination with a construct in which the input gene is flanked by homology regions matching the target site. The precise insertion of transgenes in this manner eliminates the risks associated with random insertion (position effects, interference with/mutation of endogenous genes) and can be used to generate either cisgenic or transgenic plants.

Sequence-specific nucleases can also simplify breeding strategies by facilitating trait stacking to optimize crop performance (e.g., herbicide tolerance and pest resistance in the same variety). As the number of genes to be incorporated in the germplasm increases, the associated breeding programs become more complex. It is therefore desirable to integrate multiple transgenes at defined loci to facilitate subsequent breeding efforts. This has been addressed by equipping transgene cassettes with unique sequences that act as landing pads for ZFNs in subsequent rounds of editing. If the first transgene carries a unique sequence corresponding to a ZFN target site, then the transgenic event can be super-transformed with another transgene and a suitable ZFN construct such that the incoming transgene is added in tandem to the incumbent transgene. If the second transgene also carries a unique landing pad (different to the first), the process can be repeated any number of times. Using this approach, the herbicide resistance gene *aad1* coding for aryloxyalkanoate dioxygenase has been integrated adjacent to an existing phosphinothricin acetyltransferase (*pat*) transgene in maize.

In an ingenious approach, the targeted integration of 7.1-kb and 16.2-kb multigene constructs was demonstrated in soybean embryos. The constructs were delivered as circular donor DNAs together with ZFN constructs targeting both the endogenous fatty acid desaturase gene *FAD2-1a* and the donor DNA. Following cleavage by the ZFNs, microhomology between the compatible overhangs in the target and the donor promoted the seamless insertion of the multigene constructs into the *FAD2-1a* gene by NHEJ. By simultaneous cleavage at the 5j and 3j ends of a target gene and the delivery of an appropriately designed donor template, gene exchange can also be achieved by HDR. For example, ZFNs have been used to knock out the gene encoding inositol-1,3,4,5,6-pentakisphosphate 2-kinase (IPK) in maize following this strategy. The IPK enzyme catalyzes the final step in the pathway leading to the synthesis of phytate, the main phosphate storage molecule in seeds and grains. To achieve gene inactivation, IPK-specific ZFNs and a donor DNA encoding the marker gene *pat* flanked by homology arms were co-delivered, and the *IPK1* gene was disrupted by the integration of *pat*, enabling the identification of targeted events by screening seedlings for bialaphos resistance. Using a similar approach, CRISPR/Cas9 has been used to replace the weak endogenous maize *ARGOS8* promoter with a more active promoter sequence, producing varieties with increased grain yield under drought stress conditions in the field.

Delivery Methods

When deciding which of the three major genome-editing tools to use, one major issue is how and in what form the tool will be delivered to the plant cell. For most plant species, more than one delivery method can be used, and the genome-editing tools can be supplied as DNA, RNA, proteins, or RNPs. There are certain technique-specific limitations regarding the choice of material: ODM requires the use of oligonucleotides, and only CRISPR/Cas9 and its derivatives feature an RNA component allowing RNP delivery.

DNA

Most, if not all, methods for the delivery of genome-editing reagents have evolved from methods originally developed for transgenesis, which involved the delivery of recombinant DNA. Here researchers have their pick and can select the best-established method in the plant species of interest or the method best suited for the experiment at hand. The methods discussed in this part are suitable for the delivery of constructs encoding programmable sequence-specific nucleases and base editors. Oligonucleotides for ODM also comprise or contain DNA.

Agrobacterium-mediated Transformation

The soil-dwelling plant pathogen *Agrobacterium tumefaciens* has the natural ability to transfer a segment of DNA (T-DNA) from a resident plasmid into plant cells and subsequently facilitate the integration of that DNA into the nuclear genome, resulting in the development of plant tumors (crown galls) that shelter the bacteria and produce metabolites they can use as an energy source. By replacing the bacterial genes normally found on the T-DNA, this method can be used to introduce transgenes into any plants in the Agrobacterium host range. To streamline this process in the laboratory, several disarmed binary vector systems have been developed. Disarmed in this context means that the genes that induce tumor growth and metabolic activity are removed, and binary means that the T-DNA is located on a small shuttle vector to facilitate cloning in *Escherichia coli* and maintenance in Agrobacterium, whereas the virulence genes needed for DNA transfer are located on a second helper plasmid.

T-DNA transfer requires the bacterial and plant cells to be brought into close contact, which can be achieved by syringe or vacuum infiltration of leaves, the co-cultivation of Agrobacterium and suitable plant tissues, cells, or protoplasts, by floral dip or by spraying. However, the efficiency of these methods is species-dependent: The infiltration of leaves does not work well in monocots, floral dip is used almost exclusively with Arabidopsis, and the most competent tissue for transformation also differs by species and, in some cases, by cultivar. Regardless of the transfer method, Agrobacterium can routinely introduce large T-DNAs of ~25 kb, which is usually sufficient to carry the *S. pyogenes cas9* gene (~4 kb), several gRNA genes, and the primary transgene of interest. However, binary plasmids larger than 10–15 kb are difficult to maintain in standard *E. coli* strains. One advantage of Agrobacterium-mediated transformation is that the quality of integration events is often high: The T-DNA is usually intact and there is a high frequency of single-copy/low-copy-number integration events. The main drawback is the host specificity of available Agrobacterium strains: Dicots are generally much more amenable than monocots, although the transformation of recalcitrant monocots is possible using specialized techniques to increase competence.

In the context of genome editing, Agrobacterium is used most often to generate transgenic plants that have one or more T-DNA copies stably integrated into the genome, allowing the constitutive or inducible expression of a primary transgene along with a selectable marker. Efficient transformation and regeneration protocols exist for many plant species, making this approach particularly suitable for the delivery of DNA constructs encoding the genome-editing reagents and the subsequent regeneration of transgenic plants. Interestingly, it is also possible to use Agrobacterium to transiently express the CRISPR/Cas9 components in tobacco leaves, which allows the recovery of non-transgenic genome-edited plants: 17% of the genome-edited plants recovered using this method were non-transgenic. However, both the stable transformation and transient expression approaches come with a trade-off between on-target and off-target mutations. Stable transformation leads to the sustained expression of the genome-editing components (e.g., Cas9 and gRNA in the CRISPR/Cas9 system), which increases the efficiency of on-target mutations but also risks the accumulation of mutations at off-target sites. In contrast, transient expression achieves a lower frequency of both on-target and off-target mutations because the components are available for a shorter time, but the payoff is that no foreign DNA is integrated into the genome.

Genome-editing tools can also be delivered using deconstructed viral vectors, including those based on single-stranded DNA viruses of the Geminivirus family, or single-stranded RNA viruses such as Tobacco rattle virus, Tomato mosaic virus, and Tobacco mosaic virus. Such deconstructed virus genomes generally lack the ability to move from cell to cell and are therefore incorporated into the T-DNA and introduced into the host plant by agroinfiltration. The ability of viral vectors to replicate is particularly interesting in the context of genome editing because high copy numbers of genes or HDR templates can be generated in vivo to achieve higher editing efficiencies in combination with a targeted DSB. The advantage of RNA viruses is that the replicated RNA genome does not integrate, but it also means that they cannot be used to amplify DNA repair templates. Their limited cargo capacity is also a drawback, preventing the delivery of large genes such as *SpCas9*. Theoretically, viral vectors could be delivered to plant cells or tissues without

Agrobacterium, but for genome-editing purposes, viral vectors are routinely delivered as part of a T-DNA construct.

Particle Bombardment

Particle bombardment is a physical transformation method that is particularly suitable for species beyond the Agrobacterium host range. Tungsten or gold microparticles are coated with DNA, RNA, protein, or RNPs, and accelerated by gas pressure into cells. In the aqueous environment of the cell, the nucleic acid cargo can elute from the microparticles and integrate into the genome, or remain as an extrachromosomal construct (transient expression). The advantage of particle bombardment is its genotype independence and versatility, which means that it can be used to transform a broad range of plant species in the form of different explants. However, the quality of the integration events is usually lower than Agrobacterium-mediated transformation and rarely results in clean single-copy integration of the DNA construct. Both circular and linear DNA can be introduced into plant cells by particle bombardment, and the co-delivery of different DNA fragments or plasmids is quite efficient because they can be coated onto the same microparticles. A recent study in wheat demonstrated the ZFN-mediated precise editing of the three homoeologous *ALS* genes following bombardment with the ZFN construct and a short dsDNA donor with compatible overhangs.

Protoplast Transfection

Most techniques that are routinely used for the introduction of DNA into mammalian cells are unsuitable for intact plant cells because the cell wall is a physical barrier. However, if the cell wall is removed enzymatically, the resulting protoplasts can be transfected using techniques similar to those used with mammalian cells. The protoplasts can resynthesize their cell walls and proliferate in suspension, but the regeneration of intact plants is a severe bottleneck and can only be achieved routinely in a few model species. Nevertheless, protoplasts are useful for testing the efficiency of different genome-editing reagents, at least in species that readily allow protoplast isolation. Although there are many different transfection reagents for mammalian cells, the transfection of plant protoplasts is usually achieved using polyethylene glycol (PEG), which allows the efficient introduction of both circular and linear DNA by permeabilizing the plasma membrane. Specialized vectors are not required because the transfer mechanism is physical rather than biological, and multiple plasmids can therefore be introduced simultaneously. The introduced DNA can be expressed transiently or it may integrate into the genome, resulting in high frequencies of stable transformation. For example, Nicolia et al. delivered a TALEN-expressing plasmid into potato protoplasts with a transformation efficiency of ~39% and used PCR to screen for transgene-free genome-edited callus. They targeted the *ALS* gene and found 11%–13% of callus clones and 10% of regenerated shoots were mutated at the target site.

Electroporation

The electroporation of protoplasts involves the establishment of a strong electric field that changes the permeability of the cell membrane by opening transient pores, allowing the diffusion of nucleic acids. Electroporation has been used for DNA delivery to the protoplasts of various crop species, and for gene targeting in tobacco BY-2 cells. Protoplasts are not a popular explant for stable transformation because they are challenging to generate and even more challenging to regenerate, but they are attractive for the development of high-throughput genome-editing approaches because electroporation can be automated. Electroporation of microspores and some types of intact plant cells is also possible. Wheat microspores have been transformed by electroporation with a plasmid carrying *Cas9* and gRNA genes, but both the transfection efficiency and the genome editing efficiency were low.

Other Delivery Methods

Plant cells or embryos can be mixed in a DNA solution containing a suspension of needle-like silicon carbide whiskers, which create holes in the plant cell that allow the uptake of DNA. This method has been used for the transformation of maize but it is not an efficient method and it has not, to our knowledge, been used for genome editing. If target cells or protoplasts can be immobilized, DNA can be delivered to individual cells via thin glass capillaries under a microscope without cell damage. Microinjection requires expensive equipment and is very time-consuming because each cell must be transformed manually. Genome-editing components could theoretically be delivered by microinjection, but regeneration from single transformed cells is not possible in many plant species.

RNA

The delivery of genome-editing tools into plant cells as RNA is an interesting approach that has the advantage that RNA has a limited lifespan but can still produce many copies of an encoded protein, such as a sequence-specific nuclease. In the CRISPR/Cas9 system, the gRNA can also be introduced directly as RNA. In ODM, it is necessary to introduce ssODNs or heavily modified DNA/RNA hybrids into the cell, and the principles are similar to those guiding the delivery of RNA. Their delivery is more akin to RNA than DNA and, therefore, is included here.

Particle Bombardment

One of the main drawbacks of particle bombardment for DNA delivery is the frequent low quality of the integration events. This is overcome when using RNA, because no integration event takes place. RNA is more fragile and susceptible to degradation than DNA, particularly plasmid DNA. Nevertheless, it can successfully be delivered into plant cells by particle bombardment.

Zhang et al., for example, bombarded immature wheat embryos with plasmid DNA or in vitro transcribed mRNA representing the CRISPR/Cas9 components and regenerated wheat plants without selection. As anticipated given the higher molecular stability of DNA and the amplification effect of transcription, the DNA-based approach resulted in a higher percentage of mutated plants (3.3%) than the delivery of RNA (1.1%), but a higher percentage of the mutants were homozygous following RNA delivery (35%) compared to DNA (27%). Two of seven homozygous mutants generated by bombardment with DNA were transgene free, compared to all six homozygous mutants generated with RNA (reflecting the inability of RNA to integrate). Particle bombardment has also been used for ODM in tobacco, maize, rice, and wheat in order to generate herbicide-resistant plants by targeting the *ALS* gene.

Protoplast Transfection

Protoplast transfection can also be used to deliver RNA. Stoddard et al. compared the delivery of a TALEN pair targeting the *N. benthamiana ALS* gene as either a DNA construct or in vitro transcribed mRNA. The delivery of DNA achieved a mutation frequency of >70%, whereas mRNA delivery was significantly less efficient (~6%). Because the stability of different mRNA molecules can vary, several combinations of 3j and 5j untranslated regions were tested, which doubled the mutation frequency achieved using mRNA. Interestingly, the authors also observed different mutation signatures for DNA and mRNA delivery, even though the same TALEN pair was used: DNA delivery yielded three times as many insertions as mRNA delivery and the insertions were generally >10 bp when the DNA construct was used but <10 bp for mRNA delivery. PEG-mediated protoplast transfection with modified oligonucleotides was also used to generate the Cibus sulfonylurea-tolerant canola line discussed above and confirms that this method is also suitable for ODM.

Proteins and RNPs

The delivery of genome-editing tools, particularly sequence-specific nucleases, in the form of proteins or RNPs is becoming increasingly popular. The high efficiency of these tools allows the targeted changes in the plant genome to be induced within the short half-life of the proteins, and mutated plants can be regenerated without selection. Particle bombardment and PEG-mediated protoplast transfection are the two most commonly reported delivery method for proteins and RNPs, but electroporation, silicon carbide whiskers, and microinjection could also be used in theory.

Particle Bombardment

Svitashev et al. were the first to demonstrate that bombardment can also be used to deliver Cas9/gRNA RNPs to plant cells. They bombarded maize embryo cells with RNPs targeting four different genes and regenerated a surprisingly high proportion of mutated plants without selection (2.4%–9.7%), including 0.3%–0.9% with biallelic mutations. They also compared the on-target and off-target mutation frequencies of the same gRNA delivered as a DNA construct or as a RNP, and found that a comparable proportion of plants was mutated on target (4.0% and 3.7%, respectively) whereas 2% of the plants bombarded with DNA, but not a single plant bombarded with RNPs, had off-target mutations. These findings support earlier studies reporting substantially lower off-target mutation frequencies for RNPs compared to DNA. Efficient genome editing following the bombardment of immature wheat embryos with Cas9/gRNA RNPs has also been reported, with the recovery of 4–5 independent mutants per 100 embryos without selection.

Protoplast Transfection

The PEG-mediated transfection of lettuce (*Lactuca sativa*) protoplasts has been achieved with pre-assembled Cas9/gRNA RNPs, and intact lettuce plants were regenerated from edited callus clones. Interestingly, ~6% (2/35) of the callus clones featured monoallelic mutations and 40% (14/35) featured biallelic mutations that were passed on to the T1 generation. The DNA-free delivery of genome-editing reagents is particularly useful for plant species such as grapevine (*Vitis vinifera*) and apple (*Malus domestica*) that do not allow the outbreeding of transgenes. Accordingly, the powdery mildew susceptibility locus *MLO-7* in grapevine and three fire blight susceptibility loci in apple were targeted by transfecting protoplasts with Cas9/gRNA RNPs, and the deep sequencing of genomic DNA extracted after transfection revealed that up to 0.1% of the reads in grapevine and up to 6.7% of the reads in apple contained indels. In petunia (*Petunia* × *hybrida*), the transfection of protoplasts with RNPs achieved mutation frequencies of 5.3%–17.8% in four target genes.

Selection

The delivery of genome-editing reagents to plant cells or tissues is usually followed by a tiered process of selection and/or screening to identify the small proportion of transformants, the genome-edited events, and among the latter, the events in which the mutations are on-target and have the desired effect. In contrast to conventional plant transformation, genome-editing experiments have two separate outcomes: The presence of the construct expressing the genome-editing tools and the DNA modification itself. Either or both outcomes can be desirable and the applicable screening/selection strategy differs in each case. An efficient selection strategy is particularly important when homologous recombination events must be identified and a double selection strategy may be required to avoid random integration of the donor sequence elsewhere in the genome.

The most appropriate selection strategy depends on the plant species and delivery method. Selectable marker genes that confer antibiotic or herbicide resistance are often used, as well as visual markers and selectable phenotypes. However, most of these strategies require the transfer of a heterologous marker gene into the plant, either positioned alongside the genome-editing components on the same vector, or delivered on a separate plasmid. Although such markers can, in many cases, be removed by outcrossing, the use of marker genes may be undesirable if the new crop variety is intended for commercial exploitation due to the diverse country-dependent definitions and regulations covering genetically modified organisms (GMOs). As an alternative, several DNA-free screening and selection strategies are now available that do not rely on the introduction of marker genes.

Marker genes that allow positive selection (selection for the *presence* of the gene) are an important tool in plant breeding because they allow the propagation of transformed plant cells while repressing the growth of the (usually much more abundant) untransformed cells. This is particularly valuable for species and genotypes in which transformation is inefficient. Marker genes that allow negative selection (selection for the *absence* of the gene) are used less often, but are ideal for the identification of homologous recombination events and marker-free plants. Some marker genes allow both positive and negative selection depending on the selection agent. For example, the *dao1* gene encodes a D-amino acid oxidase (DAAO), which can convert certain toxic amino acids into nontoxic ones, allowing positive selection, and can also convert certain nontoxic amino acids into toxic ones, allowing negative selection.

In genome-editing applications, positive selectable markers are often introduced along with the other components to allow the selection of cells that are stably transformed with the genome-editing construct. Antibiotic resistance genes are used for this purpose and these are often sourced from bacteria, such as the *E. coli* genes *nptII* (neomycin phosphotransferase, conferring resistance to neomycin and kanamycin) and *hpt* (hygromycin phosphotransferase, conferring resistance to hygromycin). Whereas antibiotic resistance is usually only needed for the initial selection of transformants and then becomes superfluous, herbicide resistance markers serve a dual purpose because they allow initial selection in the laboratory but also subsequent selective propagation in the field, and on that basis, they may even provide a desirable trait in the final crop. Genetically modified crops often carry markers that confer resistance to phosphino-thricin (*bar*, *pat*), glyphosate (*EPSP synthase*), or sulfonylurea herbicides (*acetolac-tate synthase*). Although antibiotics, herbicides, and other toxic substances are well established for the selection of transformed plant cells, it should be noted that they impose abiotic stress on the cells and can hamper their growth.

The preferred selection marker depends on the plant species and genotype. In potato, the *nptII* gene is used almost exclusively for selection because kanamycin resistance allows the recovery of transformed potato plants more quickly than either hygromycin or phosphinothricin. In contrast, kanamycin resistance is rarely used for selection in rice because regeneration is inefficient and the callus tissue appears to have a degree of natural resistance, whereas the *hpt* gene is favored because hygromycin selection distinguishes between untransformed and transformed rice cells more effectively. Hygromycin is also commonly used in barley (*Hordeum vulgare*), because it prevents the regeneration of untransformed plants effectively. For wheat, a species notoriously recalcitrant to transformation, selection works well with either *bar* or *hpt*. Several antibiotic and herbicide resistance genes have been used for the selection of transformed maize plants, but *pat* and *bar*—conveying resistance to phosphinothricin herbicides—are used most often for the development of commercial maize varieties. Due to concerns regarding the use of microbial antibiotic or herbicide resistance genes, alternative selection strategies are being developed based on plant-derived genes or non-antibiotic/herbicide selection systems. An example of the latter is the phosphomannose isomerase/mannose selection system, which exploits the fact that mannose is not toxic, but is naturally phosphorylated by plants and inhibits glycolysis. The enzyme phosphomannose isomerase catalyzes the interconversion of mannose-6-phosphate to fructose-6-phosphate and, thus, makes mannose available to the metabolism of transformed cells, allowing the efficient use of this system for selection.

Visual/Screenable Phenotype

Reporter genes that confer a visible or screenable phenotype are usually non-selectable (i.e., they do not promote or compromise the health of the plant alone or in the presence of a selection agent), but their activity allows transformed and untransformed plants to be identified visually. The visual output can be based on an enzymatic reaction that requires an appropriate substrate (e.g., β-galactosidase, β-glucuronidase, or luciferase), the inherent fluorescence of a protein such as green fluorescent protein (GFP) and its derivatives, or a screenable phenotype caused by an introduced transgene (e.g., anthocyanin production), or by changes mediated by the genome-editing tool itself (e.g., a drooping leaf phenotype in rice or a seedless fruit phenotype in tomato]). A popular target gene to test the efficiency of sequence-specific nucleases and different delivery methods is phytoene desaturase (*pds*), because a homozygous knockout results in an albino phenotype. The *pds* gene has been targeted in numerous species including rice, wheat, poplar (*Populus tomentosa* Carr.), cassava, and apple. However, because the albino plantlets lack chlorophyll and, thus, cannot survive in soil, these mutants are mostly of interest for testing and establishing genome-editing protocols. This issue could be addressed by the simultaneous mutation of the primary target gene and an endogenous gene conferring a screenable phenotype that does not impair plant growth.

The emergence of DNA-free methods to edit or engineer plant genomes requires alternative strategies for the identification of edited events without the use of selectable or screenable markers. One approach is high-throughput screening based directly on DNA analysis. The CRISPR/Cas9 components were delivered as DNA or RNA into two hexaploid bread wheat and two tetraploid durum wheat varieties, and plants were regenerated en masse without selection. To identify on-target mutants, a pooled screening approach was used to simultaneously test 3–4 plants regenerated from the same immature embryo in a PCR restriction enzyme assay. Pools that tested positive for mutations were deconvoluted and the plants were tested one by one. Across 11 experiments covering four wheat varieties and seven target genes, the authors identified 1.0%–9.5% mutated wheat plants regenerated from 800–1600 bombarded embryos. Of the T0 mutant plants, 44%–100% were transgene free. It should be noted that many wild-type plants will regenerate alongside the desired mutant plants in the absence of selection. For mutants without a phenotype, this increases the number of plants that must be screened regardless of the efficiency of the employed sequence-specific nuclease. Existing screening methods for marker-free plants regenerated without selection have been developed for marker-free transformation approaches, but cannot necessarily be adapted to screen genome-edited plants generated using DNA-free methods. Marker-free plants are still transgenic and PCR screening can confirm the presence or absence of the transgene, but the small insertions and deletions generated by Cas9 activity are not as easy to identify and some form of sequencing is often required.

Regeneration

The final, crucial step in any plant transformation protocol is the successful in vitro regeneration of whole plants from transformed plant cells or tissues. For many species and genotypes, the regeneration of transformed plants is a bottleneck because regeneration frequencies are low and the necessary tissue culture steps are time-consuming. There are two main pathways for the regeneration of plants from transformed cells and tissues: Organogenesis (the induction of shoots on the transformed explant and subsequent rhizogenesis, essentially the re-initiation of post-embryonic development) and somatic embryogenesis (the reprogramming of transformed somatic cells to develop into embryos, recapitulating the entire developmental pathway). Both processes are highly dependent on external stimuli provided by plant hormones (auxins and cytokinins) and exploit the totipotency of individual plant cells, i.e., their ability to change their normal developmental fate and generate all the cells and tissues found in whole plants. However, despite the growing body of knowledge concerning the factors that influence regeneration, the interplay is so complex that the optimization of regeneration protocols is still often achieved by trial and error. Insights gained from one species are not necessarily transferrable to other species or even other genotypes of the same species. Protocols for one crop variety can, therefore, only serve as starting point for a related variety. Monocots are usually more challenging to regenerate than dicots, and the undifferentiated monocot meristem cells that can be used for regeneration are thought to be particularly recalcitrant to Agrobacterium-mediated transformation. An interesting alternative approach that sidesteps the need for regeneration was recently described by two research groups using maize haploid inducer lines with CRISPR/Cas9 cassettes for genome editing in elite inbred maize and wheat lines. Following haploid induction, the chromosomes of the haploid inducer parent were eliminated and only the edited, elite chromosomes remained. Such techniques are particularly attractive because they enable direct germline editing, avoid tissue culture, and are potentially genotype-independent.

Suitable Tissues

The plant tissues used for transformation must not only tolerate the stressful transformation process, but must also contain cells that can dedifferentiate into callus and then undergo organogenesis or somatic embryogenesis to regenerate into whole plants. Embryonic tissues are preferred targets in most crops because their cells have a higher regeneration capacity than the differentiated cells of older tissues.

Maize is one of the few monocots that can be transformed rather efficiently by Agrobacterium and particle bombardment (up to 50% efficiency in the presence of a selection marker). Initial maize transformation protocols were based on protoplasts because some maize genotypes can be regenerated from protoplasts into whole plants. However, the lack of efficient and genotype-independent regeneration protocols means that protoplasts are rarely used any more. Most current transformation protocols use immature embryo tissues. Svitashev et al. demonstrated a DNA-free genome-editing approach by delivering pre-assembled RNPs into immature maize embryos by particle bombardment, and were able to regenerate mutated plants without selection. Rice transformation is also quite efficient nowadays, with 50%–90% of callus clones yielding transformants even though rice was initially considered to be a recalcitrant crop. Many rice genotypes can be transformed at least an order of magnitude more efficiently than other cereals. Agrobacterium-mediated transformation is used in most cases, but particle bombardment is frequently used in japonica rice cultivars. Immature embryo tissue can be transformed efficiently, but other explants such as scutellum-derived callus have also been used successfully with Agrobacterium. Rice protoplasts are often used to test the efficiency of CRISPR/Cas9 components, but the regeneration of whole plants is still challenging and far from routine.

The genetic transformation of wheat was achieved most recently among the major crops. Particle bombardment is the most popular transformation method because it is more efficient and works in more varieties than Agrobacterium, and the latter can trigger cell death in immature wheat embryos. Regeneration of transformed wheat works best with immature embryo material (the immature inflorescence and the scutellum), but mature embryos are more convenient because they are available throughout the year and the establishment of efficient transformation methods is therefore a key research area. Particle bombardment has been used to deliver Cas9/gRNA RNPs into immature wheat embryos, and mutated plants were regenerated with a frequency of >4% relative to the number of transformed callus clones. There are no reliable protocols for the regeneration of plants from wheat protoplasts, so in the context of genome editing, protoplasts are mostly used for testing purposes. Microspores are advantageous targets for genome editing because they are haploid and, therefore, carry only half the normal number of target alleles, but neither transformation nor regeneration are very efficient yet.

The transformation of barley is usually achieved by mixing immature embryos with Agrobacterium. The barley variety Golden Promise can be transformed with up to 25% efficiency in the presence of a selection marker, but other varieties present more of a challenge. The Agrobacterium-mediated transformation of barley ovules is thought to be genotype-independent, but ovule isolation is laborious and requires skilled personnel. Efficient genome editing using the CRISPR/Cas9 system has been demonstrated in barley using both Agrobacterium and particle bombardment.

Sorghum (*Sorghum bicolor*) is difficult to transform and regenerate. Immature embryos are almost exclusively used as explants for transformation, but despite the optimization of cultivation and transformation conditions, overall efficiencies remain below 20% for particle bombardment and below 10% for Agrobacterium-mediated transformation even in the presence of a selection marker. Using an improved transformation protocol based on an Agrobacterium ternary vector system and the presence of a surfactant, regeneration efficiencies increased to 6%–29% in the popular Tx430 sorghum variety and three African cultivars, allowing the knockout of the centromere-specific histone H3 gene (*Sb-CENH3*) using the CRISPR/Cas9 system.

Potato is transformed efficiently by Agrobacterium although transformation and/or regeneration remain a bottleneck in some varieties. Popular explants for transformation include leaves from plants cultured in vitro, stem internode segments, and minituber segments. In contrast to most crop species, potato protoplasts can be readily regenerated into plants and have been used for the transient expression of DNA-encoded TALENs or Cas9+gRNA and the DNA-free delivery of Cas9/gRNA RNPs, followed in each case by the regeneration of mutated shoots. DNA-free transformation is particularly useful in potato (a tetraploid crop with complex inheritance behavior) because outcrossing the integrated transgene or T-DNA while keeping the desired genetic context can be very difficult.

Soybean can be routinely transformed using either Agrobacterium (preferably with cotyledonary node explants) or particle bombardment (preferably with meristems), but the most efficient transformation protocols are not genotype-independent. A variety of other explant tissues have been tested to identify the best targets for transformation and regeneration. Interestingly, two additional low-tech transformation methods have been reported for soybean that deliver DNA directly into the flower, i.e., pollen tube pathway transformation and ovary-drip transformation, but it is not clear whether these methods are reproducible. The transformation efficiencies in some soybean genotypes are low and this hampers genome editing and high-throughput genomic analysis]. ZFNs, TALENs, and the CRISPR/Cas9 system have been used to generate mutations in soybean.

Cassava is usually transformed using Agrobacterium with friable embryonic callus as the explant of choice, but despite protocol improvements, it still takes 20–30 weeks to regenerate transgenic shoots and the transformation and regeneration frequencies both remain low. Cotyledons are an alternative explant, but the transformation and regeneration frequencies are even lower. Because both genetic transformation and conventional breeding are challenging in cassava, genome editing is of particular interest. Odipio et al. used CRISPR/Cas9 to knock out *pds* in cassava and achieved an unanticipated efficiency of 93%–95% regenerated plants with the associated albino phenotype. In another study, two genes involved in the interaction with a cassava virus were knocked out separately or simultaneously, and even though only 6–15 plants were regenerated per construct, 78% carried biallelic or homozygous mutations in the corresponding target gene.

These studies demonstrate that, despite unresolved challenges in the transformation and regeneration of some species, genome editing is a valuable breeding tool because of its high efficiency. This ensures that even small numbers of plants are sufficient to generate the desired mutants.

Timelines

One of the main hurdles preventing the establishment of high-throughput genome-editing protocols for crops is the long overall duration of the transformation and regeneration process, which begins with the cultivation of plants to provide explant material and ends with the harvesting of mature seeds from the transformed T0 generation. The most time-consuming steps are typically the in vitro selection and regeneration of plantlets and the maturation of the regenerated plants. In sorghum, the process takes 9–12 months, in maize it takes ~9 months, and in wheat—depending on the variety and transformation method—it takes 7–10 months. Transformation and regeneration are relatively efficient in rice but the process still takes ~7 months. In soybean, it takes 7–9 months when particle bombardment is used for transformation. Efforts to reduce the time required for in vitro cultivation (or to abolish this phase all together) could shorten the overall timelines for the transformation of important crops. Furthermore, increasing the photoperiod of spring wheat, durum wheat, barley, chickpea (*Cicer arietinum*), and pea (*Pisum sativum*) plants to 22 h was recently shown to facilitate the cultivation of six generations in 12 months rather than the usual two or three generations.

Regulatory Aspects of Plant Genome Editing

The regulatory framework that covers genome-edited crops in different countries has a major impact on their development and marketability. In most countries, the current biosafety framework was introduced in the 1980s and 1990s to regulate GMOs generated by conventional gene transfer, and there are ongoing discussions to determine whether the rules are also suitable to regulate products generated by genome editing. Thus far, only Argentina and Brazil have passed supplementary legislation to specifically address the regulatory issues associated with genome-editing applications.

There are two major regulatory triggers for GMOs, one based on the process and the other based on the product. The European Union has a process-based regulatory trigger, whereas Canada has a product-oriented evaluation system for plants with novel traits. The USA uses a hybrid system where the regulatory trigger is process-based but the risk assessment is product-based, and genome-edited plants are reviewed on a case-by-case basis.

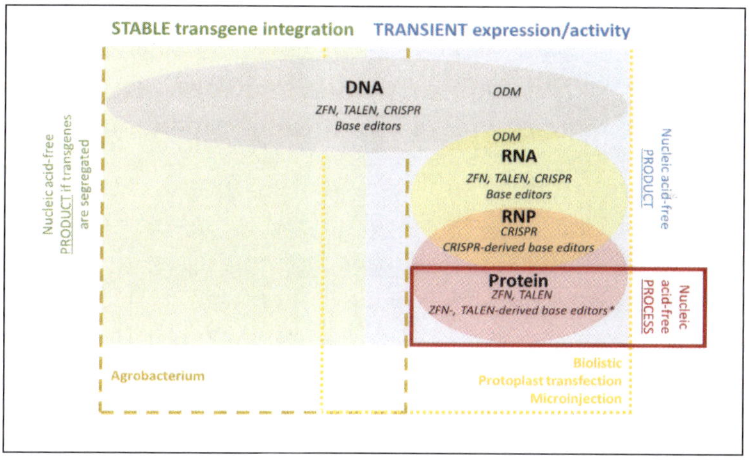

Representation of the relationships between genome-editing tools, delivery methods, and outcomes. The figure shows which tools can be delivered as DNA, RNA, RNPs, or proteins, and which delivery methods are suitable for each cargo type. It also indicates which cargo and delivery methods are available for stable transformation and transient expression, and categorizes them according to the use of nucleic acids (which is relevant for the regulatory assessment of generated plants). The sizes of the shapes are intended to promote visual clarity and do not indicate any relative importance among the methods.

The same genome-edited plant can, therefore, be classified differently from country to country depending on the regulatory framework, making it either subject to or exempt from the rigorous assessment and authorization procedures for GMOs. These aspects must be taken into account when deciding which genome-editing technology to use and how to deliver it. In Argentina, for example, applications for plants that do not retain the transgene in the final product ("null segregants") are not regulated as GMOs. There are several examples of genome-edited crops in which Agrobacterium-delivered transgenes have been removed by segregation in subsequent generations resulting in transgene-free lines, including the segregation of TALEN-encoding genes and a selection marker from rice with a fragrant phenotype, and the segregation of CRISPR components from virus-resistant cucumber plants and powdery mildew-resistant tomato.

In process-triggered systems, administrative, legislative, or court decisions are necessary to clarify which genetic modification applications fall under the legislation/GMO definition. On 25 July 2018, the European Court of Justice ruled that organisms developed using genome-editing technology fall under the obligations of Directive 2001/18/EC. This means that genome-edited crops must go through a lengthy and costly comprehensive assessment of indirect and long-term effects, and that they (and food and feed derived from them) must fulfill the requirements for detection, identification, and quantification. In plants with single-nucleotide changes or short indels, it is not possible to determine whether the mutations occurred naturally, were randomly introduced by classical mutagenesis, or were deliberately introduced by genome editing. The market surveillance of seeds or food and feed products will be a serious challenge for international trade between countries with different GMO regulations.

Genome editing in plants became possible when the first designer nucleases were described in the late 1990s, but the field has exploded in popularity since 2012 following the development of the CRISPR/Cas9 system. Depending on the desired outcome, editing can be achieved using sequence-specific nucleases, ODM, or base editors. The easiest approach, and therefore the most widely reported, is the use of sequence-specific nucleases to disrupt genes and generate knockout mutations. However, the range of qualitative traits of agricultural interest that can be engineered by gene disruption is rather limited. For many traits, especially quantitative traits, the modulation of gene function or expression is more desirable. This has been achieved by converting sequence-specific nucleases into transcriptional regulators, and by the direct editing of cis-acting regulatory sites within genes, as recently demonstrated in tomato. Although this approach requires more development and screening efforts to identify and evaluate the desired phenotypes, the modification of non-coding sequences and the engineering of quantitative trait variation are likely to be the next steps in genome-editing applications for commercial crops.

In addition to the mutation of DNA sequences, the properties of plants could also be modified by epigenome editing, which involves the control of gene expression by targeting DNA methylation or histone acetylation. It could have important applications in agriculture given that no genetic modification is involved and it would not be subjected to GMO regulations, at least in product-based regulatory frameworks. Epigenome editing could be used not only to modulate the expression of specific genes or gene families, but also to influence the chromatin landscape of selected genomic regions. However, before epigenome editing can be used to produce new crop varieties, more research is needed to evaluate the stability and heritability of the epigenetic changes introduced by new epigenome-editing tools.

One of the major technical hurdles associated with genome editing in plants is the dependence on tissue culture, which dictates the explants that can be used for the delivery of genome-editing constructs or components, affects the efficiency of that delivery process, and determines whether whole plants can be regenerated. To exploit the tremendous potential of genome-editing technologies, it is necessary to study gene function on a genomic scale allowing the identification of candidate genes underlying agronomic traits. In addition, to ensure a high editing efficiency, several enzymes or guide RNAs should be tested for each gene. To achieve this, much greater effort is required to develop efficient protocols for the transformation and regeneration of diverse species and varieties, preferably in an automated high-throughput manner. Direct modification of the germline or meristematic tissues in planta would bypass the tissue culture steps and overcome the species-and variety-dependent limitations of the regeneration process. This would enormously simplify and accelerate the creation of genome-edited lines, and the first progress in this direction has already been made. Very recently, researchers at Syngenta achieved genome editing in maize by combining the delivery of the *Cas9* transgene by pollen with haploid induction.

Although genome-editing technology is already mature enough to allow the generation of edited lines for commercial use, there remains a need to continuously improve and optimize all genome-editing approaches. Now that genome editing is finally a reality, it is clear that sociopolitical rather than technological factors will restrict the widespread adoption of genome-edited crops. When genetic modification involves the introduction of heterologous DNA to confer a new function, it is clear that both the process and the product fall under GMO regulations. However, the modification of one or a few bases can be achieved by a spectrum of processes, including natural mutation, random chemical/radiation mutagenesis, and deliberate editing with or without the introduction of DNA. Given that nucleic acid-free genome-editing methods are the least likely to trigger regulatory and safety concerns, they will continue to receive significant attention. In particular, the efficient delivery of proteins and RNPs should be optimized, and effective nucleic acid-free ways to identify cells containing these reagents should be established.

Although current genome-editing technologies are far more precise than traditional random mutagenesis, the acceptance of genome-edited crops remains a major bottleneck. A limited understanding of the underlying technical aspects and outcomes, together with the common misbelief that exemption from GMO regulations, would mean the uncontrolled release of new edited crops, contributing to public suspicion and opposition in many countries, particularly in Europe. This, in turn, influences the political decisions of the authorities, even those outside Europe. Once all the technical hurdles have been tackled and a genome-edited crop is finally ready to go on the market, it must comply with the legal obligations including labeling and traceability, which differ from country to country. As it is impossible to determine the process by which point mutations have been generated, international trade and market surveillance for genome-edited products will be a significant challenge. We will be soon facing these issues in practice, so the harmonization of regulations on a global level is urgently required.

Improved Nutritional Content

Agricultural biotechnology has been used to improve the nutritional content of a variety of crops in an effort to meet the needs of an increasing population. Genetic engineering can produce crops with a higher concentration of vitamins. For example, golden rice contains three genes that allow plants to produce compounds that are converted to vitamin A in the human body. This nutritionally improved rice is designed to combat the world's leading cause of blindness—vitamin A deficiency. Similarly, the Banana 21 project has worked to improve the nutrition in bananas to combat micronutrient deficiencies in Uganda. By genetically modifying bananas to contain vitamin A and iron, Banana 21 has helped foster a solution to micronutrient deficiencies through the vessel of a staple food and major starch source in Africa. Additionally, crops can be engineered to reduce toxicity or to produce varieties with removed allergens.

Genes and Traits of Interest for Crops

Insect Resistance

One highly sought after trait is insect resistance. This trait increases a crop's resistance to bugs and allows for a higher yield. These genetically engineered crops can now produce their own Bt (*Bacillus thuringiensis*), which contains toxin-producing proteins that are non-harmful to humans. Bt corn and cotton are now commonplace, and cowpeas, sunflower, soybeans, tomatoes, tobacco, walnut, sugar cane, and rice are all being studied in relation to Bt.

Herbicide Tolerance

Weeds have proven to be an issue for farmers for thousands of years; they compete for soil nutrients, water, and sunlight and prove deadly to crops. Biotechnology has offered a solution in the form of herbicide tolerance. Chemical herbicides are sprayed directly on plants in order to kill weeds and therefore competition, and herbicide resistant crops have to the opportunity to flourish.

Disease Resistance

Often, crops are afflicted by disease spread through insects (like aphids). Spreading disease among crop plants is incredibly difficult to control and was previously only managed by completely removing the affected crop. The field of agricultural biotechnology offers a solution through genetically engineering virus resistance. Developing GE disease-resistant crops now include cassava, maize, and sweet potato.

Temperature Tolerance

Agricultural biotechnology can also provide a solution for plants in extreme temperature conditions. In order to maximize yield and prevent crop death, genes can be engineered that help to regulate cold and heat tolerance. For example, papaya trees have been genetically modified in order to be more tolerant of hot and cold conditions. Other traits include water use efficiency, nitrogen use efficiency and salt tolerance.

Quality Traits

Quality traits include increased nutritional or dietary value, improved food processing and storage, or the elimination of toxins and allergens in crop plants.

Genetic Engineering in Agriculture

Genetic engineering is a means of altering the phenotype of an organis m through the modification of a few well-characterized genes. This may include inserting a gene or group of genes that has been synthetically produced or isolated from another organism. This method is superior to gene modification from other means such as selective breeding for the reasons discussed. First, this method is usually much faster than other methods because it does not require waiting for multiple subsequent generations to be produced and subsequently bred. Secondly, genetic engineering allows the transfer of just a few genes rather than the numerous uncharacterized genes that may be transferred through artificial selecting or grafting. Third, and perhaps most promising for future research, is the potential for genes from unrelated species to be inserted into organisms. For example, genes from bacteria may be inserted into the genome of plants.

Agricultural Improvements from Gene Modification

Insect Resistance

One of the most economically successful instances of genetic engineering in agriculture is the development and use of Bt crops. Approximately 50% of cotton and 40% of corn planted in the United Stated has been genetically engineered to produce toxins from the bacterium Bacillus thuringiensis (Bt), which is commonly found in soil. These insecticidal toxins are known as Cry toxins. This naturally produced insecticide was once sprayed onto plants to kill caterpillar and beetle pests but is now produced inside the plant so that when the insects attempt to eat the crop they are killed. In the Egyptian Nile, the stem borer, rather than the beetle or caterpillar, is a major pest in the rice fields. When a gene from Bt that led to expression of Cry toxins was introduced into the rice plant, significant mortality of the stem borer pests was recorded (some 100% after 4 days), thus protecting the crops and allowing them to grow; meanwhile the plants retained their normal phenotypes.

The production of Cry toxins has been well described from the genetic level to translation via the ribosom. At the level of transcription, scientists have identified five sporulation-specific transcription factors, which are tightly regulated and include σ^H (active in pre-division cell), σ^F and σ^G (active in the forespore), and σ^E and σ^K (active in the mother cell). Various promoters will employ these transcription factors; BtI promoter (downstream) for the cry1Aa gene is active between stages II and VI of sporulation and utilizes the transcription factor σ^E. An upstream promoter region, BtII is active later in sporulation and utilizes σ^K. After transcription, transcriptional terminators contribute to increased mRNA stability resulting in a half-life that is five times greater than that of most bacterial mRNA. At the level of translation, a perfect Shine-Dalgarno sequence has been identified, encouraging quick binding of ribosomes. This has been suggested to further protect the mRNA from degradation.

MON810, a variety of genetically modified corn produced by a Missouri, US-based plant breeding company Monsanto, is perhaps the best known example of a herbicide-resistant GM crop, marketed with the trade name YieldGard. This variety expresses the Cry-1Ab insecticidal toxin under the control of enhanced 35S promoter from Cauliflower mosaic virus and incorporates the maize Hsp70 intron. The toxin binds to receptors in Lepidopteran insects' guts, causing a quick decease, and is especially effective against corn borer (Ostrinia nubilalis). There are no binding sites for Cry1Ab on the surface of mammalian intestinal cells, therefore, livestock animals and humans are not susceptible to this protein. MON810 is grown in a number of countries worldwide, including USA, Canada, Australia, and China. An ongoing debate on the biosafety of MON810 was initiated when several European countries banned this variety, claiming various threats to humans, lower animals and ecosystems. Despite a publication arguing against this decision, and a statement from the European Food Safety Authority, MON810 is being continuously withdrawn from additional EU countries, such as Poland in April 2012.

Herbicide Tolerance

Weeds present a key threat to crops as they compete for light, water, and key nutrients. Herbicides along with tilling practices are used to remove the unwanted growth, but both of these can cause damage either to the crops themselves or the soil and surrounding water systems. Herbicide resistance of key crops could lead to less tilling, thereby resulting in improved water quality and less soil erosion. Researchers have developed crops (soybean, alfalfa, cotton and corn) that are resistant to a key, non-toxic herbicide glyphosate. These plants were engineered by the insertion of a gene from Agrobacterium that codes for CP4 EPSP, a glyphosate-insensitive form of a required enzyme, 5-enolpyruvylshikimate 3-phosphate (EPSP) synthase. These crops are known as RoundUp Ready crops and are currently marketed by Monsanto, the makers of the herbicide RoundUp (glyphosate); as of 2005, these transgenic crops represented 87% of soybeans, 61% of cotton and 21% of corn planted in the United States. Newer research focuses on utilizing plastid genomes instead of nuclear genomes to produce herbicide tolerant plants.

Viral Resistance

Two specific cases, in which viral resistance was developed in plant species, provide examples of the wide range of possibilities in genetic engineering. The first example from the 1990s was in response to the papaya ringspot virus (PRSV) that had threatened to wipe out the Hawaiian papaya production multiple times in the past. Dennis Gonsalves, a local Hawaiian, engineered papaya to carry a mild strain of PRSV with a premature stop codon to prevent full expression of the coat protein. The second example demonstrates a method scientists may use to identify gene candidates for further study. Researchers noted that resistance to Karnal bunt disease, which affects wheat, is increased when the wheat is treated with jasmonic acid (JA). JA-treated wheat displayed marked increases in expression of WC2, WC3 and W CMDcystatins, which makes them good candidates for engineering wheat strands with greater resistance to the virus.

While researchers work to find ways to guard plants against pathogens two issues have plagued them; one is the short-lived term of the resistance due to a mutation in the pathogen, and another is the loss of fitness and other negative pleiotropic effects associated with genetic changed in the resistance (R) genes. Additionally, these genetic changes usually confer resistance to only one pathogen. In order to increase resistance while maintaining plant fitness, current efforts focus on a different gene target: pathogen-associated molecular pattern (PAMP) receptors. Specifically, a receptor from Arabidopsis that recognizes a highly conserved bacterial elongation factor (EF-Tu) has been introduced to tobacco and tomato plants, both leading to broad-spectrum resistance againstsome of the most common enemies of plant growth.

Nitrogen use Efficiency

Billions of dollars are spent each year on fertilizer that contains ammonia, a form of nitrogen that has already been fixed for utilization by plants. However, plants only utilize 30 - 40% of the applied nitrogen. Increasing their nitrogen use efficiency by only a few percentage points could save millions each year. Studies being carried out in Arabidopsis have identified a GNC gene, member of a GATA transcription factor family, that may play a role in nitrogen use; overexpression of this gene led to positive effects on plant growth. However, in rice, a negative pleiotropic phenotype was observed, suggesting the need for greater understanding of nitrogen fixation and mobilization pathways for more targeted gene modifications.

Biomass Increase

Many countries are searching for possible alternate sources of fuel. One possible source is biofuels, such as ethanol from corn. As such, there is a push to increase biomass productivity on lands already in use for purposes in addition to human consumption. Multiple genes and systems have been targeted for these changes, including overexpression, loss of function, and gene shuffling to produce changes in photosynthetic genes, lipid-biosynthetic enzymes, transcription factors, and other target pathways. Other researchers have focused on genes located outside the nucleus, in the chloroplasts. In tobacco chloroplasts expressing a desaturase gene taken from potato or cyanobacteria, marked increase in leaf lipids was noted.

Stress Resistance

As bleak forecasts of global warming and increasing environmental stressors continue, researchers search for ways to prepare plants for the coming onslaught. One possible target for increased stress resistance is a group of molecules called polyamines (Pas), which include spermidine (Spd), spermine (Spm) and putrescine (Put), their precursor. Pas have anti-stress effects through acid neutralization and antioxidant properties as well as membrane stabilizing ability. Plants genetically engineered to overexpress the genes coding for these molecules have shown improved stress tolerance, even to multiple stressors such as salinity, drought, and heat.

Other studies have taken a different approach by focusing on specific stressors and attempting to identify gene targets for future engineering. Leakey et al. placed soybeans under the stress of elevated CO_2 levels and studied changes in both phenotype and gene expression. Over the two years, they monitored changes in more than 37,000 RNA transcripts. Increases were mainly noted in transcripts associated with carbohydrate metabolis m and respiration. While this research did not lend itself to immediate changes in soybeans, it provided a foundation for further studies.

Healthy Crops

Genetic engineering has opened the door to possible changes that a decade ago would have read like a page from a science fiction novel. Researchers are now working to engineer crops with health benefits, including crops that grow vaccines and antibiotics. As geneticists turn more attention to non-nuclear genomes, many are beginning to see the chloroplasts as a potential machine for producing pharmaceuticals, such as vaccines and antibiotics,in addition to well-studied starches. Preliminary studies have already shown that genetic engineering can be used to accomplish this purpose, but expression levels have been quite variable. Additionally, oral vaccine experiments have been minimal and confined to small animal models.

Perhaps less lofty than growing pharmaceuticals is the goal of making our current plants more advantageous through the production of bioactive peptides or increased vitamin content. Rice, a ubiquitous source of nutrition around the world, has very few significant physiological properties, unlike soybean. In an attempt to increase the positive effects of rice consumption and improve world health, researchers have developed rice seeds that express beta-conglycinin from soybeans. Regions of this protein have bioactive properties including hypotensive, hypocholesterolemic and memory-enhancing activity, thus demonstrating the potential for this strategy to contribute to improving health in the 21st century, especially in less developed nations where rice and other starches serve as a main source of nutrition. Another example of genetic engineering targeted to rice is the so-called "golden rice", over-accumulating β-carotene, a precursor for vitamin A that has been shown to prevent or reduce occurrence of blindness, common among poorer populations of Asia. The amount of "golden rice" consumed in a typical third world diet could provide about fifteen percent of the recommended daily allowance of vitamin A, sufficient to prevent blindness.

Following the same path, efforts were launched towards increasing iron assimilation in populations largely relying on rice as staple food to fight iron-deficiency anemia. Three parallel approaches were applied, including (i) transfer of a ferritin gene from Phaseolus vulgaris into rice grains, increasing their iron content up to two-fold, (ii) transfer of a thermotolerant phytase to increase iron bioavailability by breaking down phytate, a substance naturally present in rice grains that inhibits iron uptake, and (iii) overexpression of the endogenous cysteine-rich metallothionein-like protein that is considered a major enhancer of iron absorption.

Global Implementation

Transgenic crops are currently in all stages of testing, from gene transfer and laboratory testing to confined field tests to successful commercialization. Golden rice, for example, has undergone over five years of laboratory testing and is now being field testing in both Bangladesh and at AgCenter Rice Research Station in Louisiana, USA. In both of these tests, the golden rice performed equivalent to the control seed while producing beneficial carotenoids. Another strain of golden rice (Golden IR64) demonstrated similar results in greenhouse tests in the Philippines. A variety of drought-resistant maize that contains the cold-shock protein B (CspB) from Bacillus subtilis has been trial tested in various African countries, including Tanzania, Uganda, Kenya, and Mozambique. Sixteen different drought tolerant varieties are currently undergoing National Performance Trials (NPT) in Kenya with other countries expected to follow later this year. Bt crops are some of the most widely used in the United States and their implementation is increasing around the world. In Kenya, field trials of Bt cotton are going into their third season. Preliminary results have shown a marked decrease in targeted pests. Bt crops are showing promise in Africa and India as well. In South Africa, Bt maize, soybean and cotton are currently being released for commercial production following a five-year field trial of Bt cotton, which demonstrated about 40% higher yields than their non-transgenic counterparts and resulted in a 42% decrease in pesticide spraying cost. In Burkina Faso, Bt cotton has been in field tests since 2003, demonstrating 30% higher yields and a reduction in spraying cost. In India, the commercialization of Bt cotton has resulted in doubling of the cotton production, with 50% of that increase attributed specifically to the Bt crop. On the horizon for India, Bt rice is currently in field tests and Bt brinjal (eggplant) has been approved for seed production, with field tests to begin soon thereafter. A large number of other GM crops are also currently in field trials, including sweet potatoes, banana, cowpeas, sorghum, wheat and sugarcane in Africa, cabbage, cauliflower, groundnut, okra, and tomato in India. What once seemed like a far-fetched dream is quickly becoming a reality here in America and around the world.

Biotechnology at the DNA Level

In order to produce the changes in phenotype discussed above, there are a number of different groups of genes that may be targeted. Some of these pathways are well understood and have been applied in commercial production. Others are only in the beginning stages and currently produce many undesirable pleiotropic traits, requiring more research to understand and eliminate the negative side-effects. A few of these gene targets are discussed below.

Photosynthetic Genes

Targeted mainly in an attempt to increase biomass, there are numerous photosynthetic genes that have been the subject of genetic modification. For example, plants that normally utilize A C_3 pathway have been transformed through the insertion of C_4 pathway enzymes such as phosphoenolpyruvate carboxylase (PEPC) and pyruvate orthophosphate dikinase (PPDK). Rice with these modifications had 30 to 35% higher yield. Another unique example was the use of gene shuffling in the area that codes for rubisco activase (RCA), which limits photosynthesis at high temperatures. This shuffling led to more thermostable varieties of arabidopsis that demonstrated higher rates of photosynthesis when exposed to moderate temperature stress.

Single Base Mutations

Instead of targeting an entire gene, some research has focused on slight modifications (alteration of one or two nucleotides) to produce changes in protein activity. In an attempt to increase the positive biological activity of the soybean beta-conglycinin peptide, one Threonine was replaced with Phenylalanine or Tryptophane. Prior to actually performing the change, computer modeling was utilized to insure correct folding and the modification was expressed in E. coli to check for activity. This research demonstrates that small changes in the genome, such as single nucleotide substitutions, can lead to significant changes in peptide activity and ultimately phenotype.

Polyamine Coding Genes

As discussed previously, polyamines are a group of molecules thought to be involved in stress responses. These ubiquitously present molecules have been targeted in plants to increase stress resistance. Modifications in the genes that are involved in the polyamine pathways have led to increased stress resistance to salinity, heat and cold, drought, and other stresses; a decrease in tolerance has been seen in plants that lack these polyamines or a portion of the synthetic pathway.

Non-nuclear Genes

Since a transgene integrated in the nuclear genome stays viable in the male gamete and is dispersed via the pollen, transgene out-cross incidence to wild relatives may be as high as 38% in sunflower and 50% in strawberries. Recently, there has been a negative public perception against GMO crops due to the documented transfer of transgenes to wild relatives of a few plant species. Maternal inheritance of a transgene, and thus effective elimination of transgene escape via pollen, is possible if a transgene is integrated into the chloroplast genome. Not surprisingly, therefore, plastids have recently become a popular target for exploration in genetic engineering. These present promising targets as their genomes (about 60 genes) are much smaller than the nuclear ones and can easily incorporate new genes or gene clusters. Moreover, expression level of the foreign protein in the chloroplast is enhanced several 100 fold compared to the nuclear-integrated transgene. Modification of plastid DNA was first accomplished in 1988; however, it has not yet been utilized in commercial crops at a broader scale. One successful modification was the expression of a transgene in plastid that led to herbicide resistance, including resistance to glyphosate, isoxaflutole, and sulfonylurea. Another exciting application of plastid genetics was the construction of luminescent plants by the expression of the bacterial luciferase pathways, involving six genes. This demonstrates the ability to translate bacterial proteins into plastids and the conservation between plastid and prokaryotic translation mechanisms. One limitation of using plastid-encoded herbicide resistance is their large number within the cell that makes it difficult to obtain a uniform population of plastid genomes. While challenging, plastid genetic modification has been accomplished, providing promise for future gains.

References

- Agricultural-Biotechnology, agricultural-biotechnology, resources: isaaa.org, Retrieved 26 July, 2019

- "Infographic: Crop Modification Techniques - Biology Fortified, Inc". Biology Fortified, Inc. Archived from the original on 2016-04-14. Retrieved 2016-12-05
- Plant-mutation-breeding-in-agriculture: researchgate.net, Retrieved 21 May, 2019
- "What does it take to bring a new GM product to market? #GMOFAQ". GMO FAQ. Retrieved 2016-12-05
- Transgenics-in-agriculture: researchgate.net, Retrieved 8 January, 2019

Agricultural Technology

4

- **Mechanized Agriculture**
- **E-agriculture**
- **Agricultural Machinery**
- **Agricultural Drone**
- **Irrigation Systems**
- **Hydraulic Agriculture**

The application of techniques for controlling the growth and harvesting of vegetable products is termed as agricultural technology. Some of the machines which fall under agricultural technology are agricultural robots, rollers, subsoilers, cultivators, etc. The topics elaborated in this chapter will help in gaining a better perspective about these agricultural technologies.

Mechanized Agriculture

Mechanization of agriculture and farming process connotes application of machine power to work on land, usually performed by bullocks, horses and other draught animals or by human labour.

Benefits of Mechanization of Agriculture

1. It Increases Production: Mechanization increases the rapidity and speed of work with which farming operations can be performed. According to D. R. Bomford, "The ploughman with his three-horse am controlled three- horse; power, when given a medium-sized crawler tractor controlled between 20 to 30 horse power. His output, therefore, went up in the ratio of about 8: 1."

2. It Increases Efficiency and Per Man Productivity: Mechanization raises the efficiency

of labour and enhances the farm production per worker. By its nature it reduces the quantum of labour needed to produce a unit of output. In the U.S.A., "the amount of human labour used to produce 100 bushels of wheat dropped from 320 hours in the year 1830 to 108 hours in 1900; by 1940 a new series of improvements has reduced labour requirements to 47 hours."

According to Hecht and Barton, "Before the World War I. it took, about 35 man hrs. to grow and harvest an acre of corn; 15.2 hrs. for an acre of wheat and 15.7 hrs. for an acre of oat. In 1945-48, the labour requirements were 23.7, 6.1 and 8.1 man hours respectively. The combined, effect of fewer hours and more bushels per acre has resulted in more than halving labour requirements per unit of production. The number of man-hours required in 1910-14 per 100 bushels of corn was 135, of wheat 106 and of oat 58; in 1945- 48, the corresponding figures were 67,34 and 23 respectively."

It is estimated that productivity per man on farms in U.S.A. is about four and a half times that in the U.S.S.R. In the U.S.S.R. in collective farms, production has raised labour productivity to a high level compared with the pre-revolutionary days; now labour is three times more productive there.

3. Mechanization Increases the Yield of Land Per Unit of Area: S.E. Johnson holds that "of 28 per cent increase in farm output in U.S.A., above the average of 1934-39 only about one-fourth is due to better weather, probably less than 15 per cent has resulted from expansion of crop, land acreage and the rest, about 60 per cent is largely accounted for by the fuller use of the improvements in crops, live stocks and machinery. Increase in the yield of crops, due to mechanization of farms, has been traced from 40 to 50 per cent in the case of maize; 15 to 20 per cent in Bajra and Paddy; 30 to 40 per cent in Jowar, Groundnut and Wheat."

4. Mechanization Results in Lower Cost of Work: It has been accepted by all that one of the methods of reducing unit costs is to enlarge the size c* the farms and go in for more intensive farming. It is found that the cost of production and the yields can be adjusted properly if mechanization is resorted to.

5. It Contracts the Demand for Work Animals for ploughing water lifting, harvesting, transport etc.: In actual operation, costs amount to little when machines are idle, whereas the cost of maintenance of draught animals remains the same during both periods of working and idleness, because animals have to be fed whether they are doing work or not. It is advantageous to use tractors when a great deal of work has to be done in a short time.

6. It Brings in other Improvements in Agricultural Technique: In its training come improvements in the sphere of irrigation, land reclamation and the prevention of soil erosion.

Besides, ploughing by tractor reclaims more land and thereby extends the cultivated area as the tractor smoothens hillocks, fills in depressions and gullies and eradicate deeps-rooted weeds. It also prevents soil erosion. Besides mechanical fertilization, contour bunding and terracing are done by mechanical methods with the help of self-propelled graders and terraces.

7. It Modifies Social Structure in Rural Areas: It results in a significant modification of the social structure in rural areas. It frees the farmers from much of the laborious, tedious, hard work on the farms. The pressure on land decreases and the status of the farmers improves.

8. It Leads to Commercial Agriculture: Mechanisation results in a shift from 'subsistence farming' to 'commercial agriculture. This shift occurs mainly due to the need for more land and capital to be associated with farmer in order to reap the full technological benefits.

This in its turn gives rise two tendencies:

- Gradual replacement of domestic or family by commercial methods, and
- Search for international markets for agricultural produce.

9. It Solves the Problem of Labour Shortage: In countries where human labour falls short of requirements in agriculture, use of machines can replace human and animal power.

10. It Releases Manpower for Non-Agricultural Purposes: Since the mechanisation of agriculture results in the employment of lesser number of persons on farms, surplus manpower may be available for other economic activities.

11. It Results in Better Use of Land: Mechanisation also results in better utilization of agricultural land for "the substitution of gasoline tractor for animal power means reduced demand. The use of machine energy, therefore, leads to good agricultural production, to trade many crops or saleable animal products in short, to an exchange economy and a system of land utilization in which cultivator rests on a different and infinitely more complex basis than is found in the local self-sufficient economy."

12. It Increases Farm Income: With the introduction of mechanisation the farm income as well as the individual income goes up. E. G. Nourse writes, "It accounts for the unparalleled rise of national income and with it the standard of living, it builds cities, it raises an ever loftier superstructure of financial, commercial and other cultural institutions; it turns loose economic agglomerates into social economies to closely knit by a thousand lines of interdependence. It creates much of the capital surplus on which modern economic progress is largely based. It constitutes, the lion's share to the public funds which support education, health and law and order. In short, not only do machine industry, and mechanisation and science render agriculture efficient, they create the very world in which this efficient agriculture can sell its bountiful crops."

13. It Reduces Fodder Area and Enlarges Food Area: "With the introduction of mechanisation in agriculture the surplus animal power would be reduced so that large areas of land required for producing fodder for it can be utilised for producing food for human consumption. The remaining cattle population would be better attended to and better fed under mechanised agriculture, for new and nourishing varieties of feeding stuff would be grown in cultural (waste lands after reclaiming them for cultivation."

Agricultural Robot

An agricultural robot is a robot deployed for agricultural purposes. The main area of application of robots in agriculture today is at the harvesting stage. Emerging applications of robots or drones in agriculture include weed control, cloud seeding, planting seeds, harvesting, environmental monitoring and soil analysis.

Autonomous Agricultural Robot.

Fruit picking robots, driverless tractor/sprayers, and sheep shearing robots are designed to replace human labor. In most cases, a lot of factors have to be considered (e.g., the size and color of the fruit to be picked) before the commencement of a task. Robots can be used for other horticultural tasks such as pruning, weeding, spraying and monitoring. Robots can also be used in livestock applications (livestock robotics) such as automatic milking, washing and castrating. Robots like these have many benefits for the agricultural industry, including a higher quality of fresh produce, lower production costs, and a decreased need for manual labor. They can also be used to automate manual tasks, such as weed or bracken spraying, where the use of tractors and other manned vehicles is too dangerous for the operators.

Designs

Fieldwork Robot.

The mechanical design consists of an end effector, manipulator, and gripper. Several factors must be considered in the design of the manipulator, including the task, economic efficiency, and required motions. The end effector influences the market value of the fruit and the gripper's design is based on the crop that is being harvested.

End Effectors

An end effector in an agricultural robot is the device found at the end of the robotic arm, used for various agricultural operations. Several different kinds of end effectors have been developed. In an agricultural operation involving grapes in Japan, end effectors are used for harvesting, berry-thinning, spraying, and bagging. Each was designed according to the nature of the task and the shape and size of the target fruit. For instance, the end effectors used for harvesting were designed to grasp, cut, and push the bunches of grapes.

Berry thinning is another operation performed on the grapes, and is used to enhance the market value of the grapes, increase the grapes' size, and facilitate the bunching process. For berry thinning, an end effector consists of an upper, middle, and lower part. The upper part has two plates and a rubber that can open and close. The two plates compress the grapes to cut off the rachis branches and extract the bunch of grapes. The middle part contains a plate of needles, a compression spring, and another plate which has holes spread across its surface. When the two plates compress, the needles punch holes through the grapes. Next, the lower part has a cutting device which can cut the bunch to standardize its length.

For spraying, the end effector consists of a spray nozzle that is attached to a manipulator. In practice, producers want to ensure that the chemical liquid is evenly distributed across the bunch. Thus, the design allows for an even distribution of the chemical by making the nozzle to move at a constant speed while keeping distance from the target.

The final step in grape production is the bagging process. The bagging end effector is designed with a bag feeder and two mechanical fingers. In the bagging process, the bag feeder is composed of slits which continuously supply bags to the fingers in an up and down motion. While the bag is being fed to the fingers, two leaf springs that are located on the upper end of the bag hold the bag open. The bags are produced to contain the grapes in bunches. Once the bagging process is complete, the fingers open and release the bag. This shuts the leaf springs, which seals the bag and prevents it from opening again.

Gripper

The gripper is a grasping device that is used for harvesting the target crop. Design of the gripper is based on simplicity, low cost, and effectiveness. Thus, the design usually consists of two mechanical fingers that are able to move in synchrony when performing their task. Specifics of the design depend on the task that is being performed. For example, in a procedure that required plants to be cut for harvesting, the gripper was equipped with a sharp blade.

Manipulator

The manipulator allows the gripper and end effector to navigate through their environment. The manipulator consists of four-bar parallel links that maintain the gripper's position and height. The manipulator also can utilize one, two, or three pneumatic actuators. Pneumatic actuators are motors which produce linear and rotary motion by converting compressed air into energy. The pneumatic actuator is the most effective actuator for agricultural robots because of its high power-weight ratio. The most cost efficient design for the manipulator is the single actuator configuration, yet this is the least flexible option.

Development

The first development of robotics in agriculture can be dated as early as the 1920s, with research to incorporate automatic vehicle guidance into agriculture beginning to take shape. This research led to the advancements between the 1950s and 60s of autonomous agricultural vehicles. The concept was not perfect however, with the vehicles still needing a cable system to guide their path. Robots in agriculture continued to develop as technologies in other sectors began to develop as well. It was not until the 1980s, following the development of the computer, that machine vision guidance became possible.

Other developments over the years included the harvesting of oranges using a robot both in France and the US.

While robots have been incorporated in indoor industrial settings for decades, outdoor robots for the use of agriculture are considered more complex and difficult to develop.

This is due to concerns over safety, but also over the complexity of picking crops subject to different environmental factors and unpredictability.

Demand in the Market

There are concerns over the amount of labor the agricultural sector needs. With an aging population, Japan is unable to meet the demands of the agricultural labor market. Similarly, the United State currently depends on a large number of immigrant workers, but between the decrease in seasonal farmworkers and increased efforts to stop immigration by the government, they too are unable to meet the demand. Businesses are often forced to let crops rot due to an inability to pick them all by the end of the season. Additionally, there are concerns over the growing population that will need to be fed over the next years. Because of this, there is a large desire to improve agricultural machinery to make it more cost efficient and viable for continued use.

Current Applications and Trends

Much of the current research continues to work towards autonomous agricultural vehicles. This research is based on the advancements made in driver-assist systems and self-driving cars.

While robots have already been incorporated in many areas of agricultural farm work, they are still largely missing in the harvest of various crops. This has started to change as companies begin to develop robots that complete more specific tasks on the farm. The biggest concern over robots harvesting crops comes from harvesting soft crops such as strawberries which can easily be damaged or missed entirely. Despite these concerns, progress in this area is being made. According to Gary Wishnatzki, the co-founder of Harvest Croo Robotics, one of their strawberry pickers currently being tested in Florida can "pick a 25-acre field in just three days and replace a crew of about 30 farm workers". Similar progress is being made in harvesting apples, grapes, and other crops.

Another goal being set by agricultural companies involves the collection of data. There are rising concerns over the growing population and the decreasing labor available to feed them. Data collection is being developed as a way to increase productivity on farms. AgriData is currently developing new technology to do just this and help farmers better determine the best time to harvest their crops by scanning fruit trees.

Applications

Robots have many fields of application in agriculture. Some examples and prototypes of robots include the Merlin Robot Milker, Rosphere, Harvest Automation, Orange Harvester, lettuce bot, and weeder. One case of a large scale use of robots in farming is the milk bot. It is widespread among British dairy farms because of its efficiency and nonrequirement to move. According to David Gardner (chief executive of the Royal Agricultural Society of England), a robot can complete a complicated task if its repetitive and the robot is allowed to sit in a single place. Furthermore, robots that work on repetitive tasks (e.g. milking) fulfill their role to a consistent and particular standard.

Another field of application is horticulture. One horticultural application is the development of RV100 by Harvest Automation Inc. RV 100 is designed to transport potted plants in a greenhouse or outdoor setting. The functions of RV100 in handling and organizing potted plants include spacing capabilities, collection, and consolidation. The benefits of using RV100 for this task include high placement accuracy, autonomous outdoor and indoor function, and reduced production costs. Examples:

- Thorvald: An autonomous modular multi-purpose agricultural robot developed by Saga Robotics;
- Vinobot and Vinoculer;
- LSU's AgBot;
- Harvest Automation is a company founded by former iRobot employees to develop robots for greenhouses;
- Root AI has made a tomato-picking robot for use in greenhouses;
- Strawberry picking robot from Robotic Harvesting and Agrobot;
- Small Robot Company developed a range of small agricultural robots, each one being focused on a particular task (weeding, spraying, drilling holes) and controlled by an AI system;
- ecoRobotix has made a solar-powered weeding and spraying robot;
- Blue River Technology has developed a farm implement for a tractor which only sprays plants that require spraying, reducing herbicide use by 90%;
- Casmobot next generation slope mower;
- Fieldrobot Event is a competition in mobile agricultural robotics;
- HortiBot: A Plant Nursing Robot;
- Lettuce Bot: Organic Weed Elimination and Thinning of Lettuce;
- Rice planting robot developed by the Japanese National Agricultural Research Centre;
- ROS Agriculture: Open source software for agricultural robots using the Robot Operating System;

- The IBEX autonomous weed spraying robot for extreme terrain, under development;
- FarmBot, Open Source CNC Farming;
- VAE, under development by an argentinean ag-tech startup, aims to become a universal platform for multiple agricultural applications, from precision spraying to livestock handling;
- ACFR RIPPA: for spot spraying;
- ACFR SwagBot; for livestock monitoring;
- ACFR Digital Farmhand: for spraying, weeding and seeding.

Roller

A roller in a typical power farming application.

The roller is an agricultural tool used for flattening land or breaking up large clumps of soil, especially after ploughing or disc harrowing. Typically, rollers are pulled by tractors or, prior to mechanisation, a team of animals such as horses or oxen. As well as for agricultural purposes, rollers are used on cricket pitches and residential lawn areas.

Flatter land makes subsequent weed control and harvesting easier, and rolling can help to reduce moisture loss from cultivated soil. On lawns, rolling levels the land for mowing and compacts the soil surface.

Rollers may be weighted in different ways. For many uses a heavy roller is used. These may consist of one or more cylinders made of thick steel, a thinner steel cylinder filled with concrete, or a cylinder filled with water. A water-filled roller has the advantage that the water may be drained out for lighter use or for transport. In frost-prone areas a water filled roller must be drained for winter storage to avoid breakage due to the expansion for water as it turns to ice.

Designs

One-piece versus Segmented

A ridged roller comprising many segments is usually called a Cambridge roller in the United Kingdom and a cultipacker in the United States; each name originated with a manufacturer in the respective country and evolved into the regionally prevalent name for the type.

On tilled soil a one-piece roller has the disadvantage that when turning corners the outer end of the roller has to rotate much faster than the inner end, forcing one or both ends to skid. A one-piece roller turned on soft ground will skid up a heap of soil at the outer radius, leaving heaps, which is counter-productive. Rollers are often made in two or three sections to reduce this problem, and the Cambridge roller overcomes it altogether by mounting many small segments onto one axle so that they can each rotate at local ground-speed.

A 12-foot smooth roller comprising eight 1.5-foot segments.

A field after rolling with a Cambridge (or similar) roller.

Smooth versus Ridged

The surface of rollers may be smooth, or it may be textured to help break up soil or to groove the final surface to reduce scouring from rain. Each segment of a Cambridge roller has a rib around its edge for this purpose. The name cultipacker is often used for such ridged types, especially in the United States.

Uses

Farming Use

Rollers are a secondary tillage tool used for flattening land or breaking up large clumps of soil, especially after ploughing or disc harrowing. Rollers are typically pulled by tractors today. Before mechanised agriculture, a team of working animals such as horses or oxen provided the power. Animal power is still used today in some contexts, such as on Amish farms in the United States and in regions of Asia where draft oxen are still widely used.

Rollers prepare optimal seedbeds by making them as flat as is practical and moderately firmed. Flatness is important at planting because it is the only practical way to control average seed planting depth without laborious hand planting of each seed; it is not practical to follow an instruction of (for example) 1-cm planting depth if the contour of the seedbed varies by 2 cm or more between adjacent spots. This is why breaking up of even small clods/lumps, and well-leveled spreading of soil, is important at planting time.

Flatter land also makes subsequent weed control and harvesting easier. For example, in mechanical weed control, controlling cultivator tooth depth is practical only with a decently flat soil contour, and in combining, controlling combine head height is practical only with a decently flat soil contour.

Rolling is also believed to help reduce moisture loss from cultivated soil.

Ganging and Trailing

Rollers may be ganged to increase the width of each pass/swath. Rollers may be trailed after other equipment such as ploughs, disc harrows, or mowers.

Cricket Pitch Use

In cricket, rollers are used to make the pitch flat and less dangerous for batsmen.

Several size rollers have been used in the history of cricket, from light rollers that were used in the days of uncovered pitches and at some stages during the 1950s to make batting less easy, to the modern "heavy roller" universally used in top-class cricket today. Regulations permit a pitch only to be rolled at the commencement of each innings or day's play, but this has still had a massive influence on the game by eliminating the shooters that were ubiquitous on all but light soils before heavy rollers were used. Heavy rollers have sometimes been criticised for making batting too easy and for reducing the rate at which pitches dry out after rain in the cool English climate.

Lawn Use

Lawn rollers are designed to even out or firm up the lawn surface, especially in climates where heaving causes the lawn to be lumpy. Heaving may result when the ground freezes and thaws many times over winter. Where this occurs, gardeners are advised to give the lawn a light rolling with a lawn roller in the spring. Clay or wet soils should not be rolled as they become compacted.

Lawn roller.

Subsoiler

Howse brand modular subsoiler mounted to a tractor.

Modular subsoiler unit, unmounted with accessories.

A subsoiler or flat lifter is a tractor-mounted farm implement used for deep tillage, loosening and breaking up soil at depths below the levels worked by moldboard ploughs, disc harrows, or rototillers. Most such tools will break up and turn over surface soil to a depth of 15–20 cm (5.9–7.9 in), whereas a subsoiler will break up and loosen soil to twice those depths. Typically a subsoiler mounted on a compact utility tractor will reach depths of about 30 cm (12 in) and typically have only one thin blade with a sharpened tip.

The subsoiler is a tillage tool which will improve growth in all crops where soil compaction is a problem. In agriculture angled wings are used to lift and shatter the hardpan that builds up due to compaction. The design provides deep tillage, loosening soil depth is deeper than a tiller or plough is capable of reaching.

The subsoiler consists of three or more heavy vertical shanks (standards) mounted on a toolbar or frame with share bolts. They can be operated at depths of 45 to 75 cm or more. A ripper normally run 35-45cm deep. Shanks are curved and have replaceable tips. Each shank fitted with a replaceable point or foot, than the chisel plough to break through impervious layer shattering the sub-soil to a depth of 45 to 75 cm and requires 60 to 100 hp to pull a single subsoil point through a hard soil. Subsoiling is a slow operation and requires high power input. The shanks should be inclined to the vertical at an angle greater than 25-300, preferably 450, and it is advisable that the height be adjustable. The points of the shanks are normally about 30 cm wide and should be easy to replace. The conditions of the point is very important and often the subsoiler fails to give good results due to the conditions of its points. Point can be fitted with horizontal wings, about 300 mm wide, which considerably increases the width of soil below ploughing depth loosened by the subsoiler. These plows are sometimes equipped with a torpedo-shaped attachment for making subsurface drainage channels. The subsoilers are raised and lowered hydraulically. Some models feature power-take-off (PTO) driven vibrating devices. The typical spacing is 76 to 100 cm between shanks. Shanks should be able to reach 2,5 to 5 cm below the deepest compacted layer. Shank spacing and height should be adjustable in the field. Towed subsoilers should have gauge wheels to control the shank's depth.

Shanks usually are from 2 to 4 cm thick. Thinner shanks are suited for agricultural use. Thicker shanks hold up better in rocky conditions, but require larger, more powerful equipment to pull them and disturb the surface more.

Various manufacturers' brochures claim that crops perform well during hot and dry seasons because roots penetrate soil layers deeper to reach moisture and nutrients. Brochures further claim that in wet conditions, the water passes more easily through the shattered areas, reducing the possibility of crops drowning.

Agricultural subsoiler implements will have multiple deeper reaching blades; each blade is called a scarifier or shank. Purdue University's Deptartment of Agriculture indicates that common subsoilers for agricultural use are available with three, five or seven shanks. Subsoilers can be up to 15 feet (4.6 m) wide, some models are towed behind tractors while others are mounted to the three-point hitch.

One type of subsoiler has a torpedo-shaped tip and is called a mole plough because the tip describes a path much like the burrow that a mole creates. Mole ploughs are used to create tile drainage, with or without tiles or tile line added. A form of this implement (with a single blade), a pipe-and-cable-laying plough, is used to lay buried cables or pipes, without the need to dig a deep trench and re-fill it.

Cultivator

A cultivator is any of several types of farm implement used for secondary tillage. One sense of the name refers to frames with teeth (also called shanks) that pierce the soil as they are dragged through it linearly. Another sense refers to machines that use rotary motion of disks or teeth to accomplish a similar result. The rotary tiller is a principal example.

F210 Honda tiller.

A tractor-mounted tiller.

Cultivators stir and pulverize the soil, either before planting (to aerate the soil and prepare a smooth, loose seedbed) or after the crop has begun growing (to kill weeds—controlled disturbance of the topsoil close to the crop plants kills the surrounding weeds by uprooting them, burying their leaves to disrupt their photosynthesis, or a combination of both). Unlike a harrow, which disturbs the entire surface of the soil, cultivators are designed to disturb the soil in careful patterns, sparing the crop plants but disrupting the weeds.

Cultivators of the toothed type are often similar in form to chisel plows, but their goals are different. Cultivator teeth work near the surface, usually for weed control, whereas chisel plow shanks work deep beneath the surface, breaking up hardpan. Consequently, cultivating also takes much less power per shank than does chisel plowing.

1949 Farmall C with C-254-A two-row cultivator.

A cultivator pulled by a tractor in Canada in 1943.

Small toothed cultivators pushed or pulled by a single person are used as garden tools for small-scale gardening, such as for the household's own use or for small market gardens. Similarly sized rotary tillers combine the functions of harrow and cultivator into one multipurpose machine.

Cultivators are usually either self-propelled or drawn as an attachment behind either a two-wheel tractor or four-wheel tractor. For two-wheel tractors they are usually rigidly fixed and powered via couplings to the tractors' transmission. For four-wheel tractors they are usually attached by means of a three-point hitch and driven by a power take-off (PTO). Drawbar hookup is also still commonly used worldwide. Draft-animal power is sometimes still used today, being somewhat common in developing nations although rare in more industrialized economies.

Industrial Use

To the extent that cultivating is done commercially today (such as in truck farming), it is usually powered by tractors, especially row-crop tractors. Industrial cultivators can vary greatly in size and shape, from 10 feet (3 m) to 80 feet (24 m) wide. Many are equipped with hydraulic wings that fold up to make road travel easier and safer. Different types are used for preparation of fields before planting, and for the control of weeds between row crops. The cultivator may be an implement trailed after the tractor via a drawbar; mounted on the three-point hitch; or mounted on a frame beneath the tractor. Active cultivator implements are driven by a power take-off shaft. While most cultivator are considered a secondary tillage implement, active cultivators are commonly used for primary tillage in lighter soils instead of plowing. The largest versions available are about 6 m (20 ft) wide, and require a tractor with an excess of 150 horsepower (110 kW) (PTO) to drive them.

Field cultivators are used to complete tillage operations in many types of arable crop fields. The main function of the field cultivator is to prepare a proper seedbed for the crop to be planted into, to bury crop residue in the soil (helping to warm the soil before planting), to control weeds, and to mix and incorporate the soil to ensure the growing crop has enough water and nutrients to grow well during the growing season. The implement has many shanks mounted on the underside of a metal frame, and small narrow rods at the rear of the machine that smooth out the soil surface for easier travel later when planting. In most field cultivators, one-to-many hydraulic cylinders raise and lower the implement and control its depth.

Row Crop Cultivators

The main function of the row crop cultivator is weed control between the rows of an established crop. Row crop cultivators are usually raised and lowered by a three-point hitch and the depth is controlled by gauge wheels.

Sometimes referred to as *sweep cultivators*, these commonly have two center blades that cut weeds from the roots near the base of the crop and turn over soil, while two rear sweeps further outward than the center blades deal with the center of the row, and can be anywhere from 1 to 36 rows wide.

Homemade sweep. Notice the inner and outer “sweep” blades.

Garden Cultivators

Small tilling equipment, used in small gardens such as household gardens and small commercial gardens, can provide both primary and secondary tillage. For example, a rotary tiller does both the “plowing” and the “harrowing”, preparing a smooth, loose seedbed. It does not provide the row-wise weed control that cultivator teeth would. For that task, there are single-person-pushable toothed cultivators.

Variants and Trademarks

Rotary tillers are a type of cultivators. Rotary tillers are popular with home gardeners who want large vegetable gardens. The garden may be tilled a few times before planting each crop. Rotary tillers may be rented from tool rental centers for single-use applications, such as when planting grass.

A small rotary hoe for domestic gardens was known by the trademark Rototiller and another, made by the Howard Group, who produced a range of rotary tillers, was known as the Rotavator.

A Japanese two-wheel tractor.

Rototiller

The small rototiller is typically propelled forward via a (1–5 horsepower or 0.8–3.5 kilowatts) petrol engine rotating the tines, and do not have powered wheels, though they may have small transport/level control wheel(s). To keep the machine from moving forward too fast, an adjustable tine is usually fixed just behind the blades so that through friction with deeper un-tilled soil, it acts as a brake, slowing the machine and allowing it to pulverize the soils. The slower a rototiller moves forward, the more soil tilth can be obtained. The operator can control the amount of friction/braking action by raising and lowering the handlebars of the tiller. Rototillers often do not have a reverse as such backwards movement towards the operator could cause serious injury. While operating, the rototiller can be pulled backwards to go over areas that were not pulverized enough, but care must be taken to ensure that the operator does not stumble and pull the rototiller on top of himself. Rototilling is much faster than manual tilling, but notoriously difficult to handle and exhausting work, especially in the heavier and higher horsepower models. If the rototiller's blades catch on unseen subsurface objects, such as tree roots and buried garbage, it can cause the rototiller to abruptly and violently move in any direction.

Rotavator

Unlike the Rototiller, the self-propelled Howard Rotavator is equipped with a gearbox and driven forward, or held back, by its wheels. The gearbox enables the forward speed to be adjusted while the rotational speed of the tines remains constant which enables the operator to easily regulate the extent to which soil is engaged. For a two-wheel tractor rotavator this greatly reduces the workload of the operator as compared to a rototiller. These rotavators are generally more heavy duty, come in higher power (4–18 horsepower or 3–13 kilowatts) with either petrol or diesel engines and can cover much more area per hour. The trademarked word "Rotavator" is one of the longest single-word palindromes in the English language.

Mini Tiller

Mini tillers are a new type of small agricultural tillers or cultivators used by farmers or homeowners. These are also known as power tillers or garden tillers. Compact, powerful and, most importantly, inexpensive, these agricultural rotary tillers are providing alternatives to four-wheel tractors and in the small farmers' fields in developing countries are more economical than four-wheel tractors.

Two-wheel Tractor

The higher power "riding" rotavators cross out of the home garden category into farming category, especially in Asia, Africa and South America, capable of preparing 1 hectare of land in 8–10 hours. These are also known as *power tillers* or *walking tractors*. Years ago they were considered only useful for rice growing areas, where they were fitted with steel cage-wheels for traction, but now the same are being used in both wetland and dryland farming all over the world. They have multiple functions with related tools for dryland or paddys, pumping, transportation, threshing, ditching, spraying pesticide. They can be used on hills, mountains, in greenhouses and orchards. Diesel designs are more popular in developing countries than gasoline.

Cultipacker

A ridged roller comprising many segments is usually called a Cambridge roller in the United Kingdom and a cultipacker in the United States.

A cultipacker is a piece of agricultural equipment that crushes dirt clods, removes air pockets, and presses down small stones, forming a smooth, firm seedbed. Where seed has been broadcast, the roller gently firms the soil around the seeds, ensuring shallow seed placement and excellent seed-to-soil contact.

The term *cultipacker* is almost exclusively applied to ridged rollers, while the terms *field roller* or *land roller* may refer to either a smooth or a ridged roller. Some farmers treat the terms as mutually exclusive, but many others treat the ridged tools as a class of field rollers. For example, C.H. Wendel's *Encyclopedia of American Farm Implements and Antiques* covers the whole category as *land rollers*. The term *cultipacker* appeared in English around 1914 and probably originated as a brand name of the C.G. Dunham Company of Berea, Ohio, which advertised "Culti-Packer" models starting around that time. That company did not have the ridged-roller subcategory to itself by any stretch, as Wendel's book demonstrates, but for whatever reason, its name for its version stuck well in many minds. By the 1920s and ever since, it has been widely used in a genericized sense, at least in some regions of the U.S. if not nationwide. In Britain, an equivalent tool is usually called a Cambridge roller or Cambridge roll; D.J. Smith's *Discovering Horse-drawn Farm Machinery* says, "The Cambridge roll, named after its makers, was a ring roller made up of numerous equally spaced rings or ridges."

Despite the suggestion of soil compaction in the "packer" part of the name, cultipackers and other land rollers exert high pressure only on the high spots (such as clods); the baseline pressure at the rest of the footprint is not especially high, which is to say, not higher than a person's footprint pressure. A person intentionally stomping hard on a particular spot can pack the dirt tighter than a cultipacker packs it. This is appropriate because seedbeds need only be firm, not excessively compacted.

An advertisement for the Culti-Packer by the C.G. Dunham Company of Berea, Ohio, USA, in the April 24, 1915, issue of *The Prairie Farmer*, a farm newspaper. The earliest mentions of cultipackers in print date from around this time period. Ads for this company show single and double versions.

E-Agriculture

E-Agriculture is an emerging field focusing on the enhancement of agricultural and rural development through improved information and communication processes. It is evolving in scope as new Information Communication Technology (ICT) applications continue to be harnessed in the agriculture sector. In this context, ICT is used as an umbrella term encompassing all ICTs including devices, networks, services and applications; these range from innovative Internet-era technologies and sensors to other pre-existing aids such as telephones, mobiles, television, radio and satellites. More specifically it involves the conceptualization, design, development, evaluation and application of innovative ways to use ICTs in the rural domain, with a primary focus on agriculture, fisheries, forests and livestock.

Mobile Agriculture

Mobile Agriculture (also shortened to mAgri) supports actors along the agriculture value chain through the use of mobile technology. Mobile technology covers a broad range of devices and the sub-categories include voice, data, network and connectivity technologies. mAgri is a subset of e-agriculture.

The introduction of mobile technology and portable, wireless devices has led to the creation of innovative services and applications that are used within the agricultural value chain in developed and developing countries. In developed markets where mechanization is more advanced and the agricultural labour force is significantly smaller than that of many developing countries, mobile agriculture applications tend to be implemented further up the value chain, for example with processors or consumers. In developing countries where a large proportion of the workforce is employed in agriculture, mobile technology is more commonly used to deliver services for producers and traders.

Applications

The GSMA's Mobile and Development Impact (MDI), an open data portal on mobile and development, has created a standard set of categories that cover applications of mobile technology in agriculture. The MDI categories were created to provide an overview of the mAgri industry and the range of mobile applications in agriculture across the world, with a focus on emerging markets. The categories listed below serve a role in the mAgri industry by providing a common framework and a standardized naming system.

Instances of the technology include:

- Market intelligence/information: Applications that use mobile technology to deliver or retrieve market information including prices;
- Trading facilities;
- Weather information;
- Peer to peer learning;
- Data collection;
- Financial services - Payments;
- Financial services - Loans;
- Financial services - Insurance;
- Learning/advisory/extension services: Applications that use mobile technology to deliver or retrieve agricultural/agronomy information and advice;

- Geospatial applications: Applications enabling data and information related to geography and space to be managed, processed, and visualized. They contribute to land and water use planning, natural resources utilization, agricultural input supply and commodity marketing, poverty and hunger mapping, etc.;
- Embedded ICT in farm equipment & processes: Applications that enable greater efficiencies in farm equipment and agricultural processes, and traceability in agricultural products' transport and marketing through mobile technologies such as RFID, wireless Internet, and cellular telephony for labelling, traceability and identity preservation:
    - Operations monitoring, quality control, and product tracking;
    - Logistics and business processes.
- Agricultural news: Applications that provide news on agriculture-related subjects.

Agricultural Market Information System

The Agricultural Market Information System (AMIS) is an inter-agency platform to enhance food market transparency and encourage international policy coordination in times of crisis. It was established at the request of the Group of Twenty (G20) in 2011. Countries participating in AMIS encompass the main producing and consuming countries of major food crops covered by the initiative: wheat, maize, rice and soybeans. AMIS is hosted by the Food and Agriculture Organization of the United Nations (FAO) in Rome/Italy and supported by a joint Secretariat, which currently (September 2016) consists of eleven international organizations and entities. Apart from FAO, these are the Group on Earth Observations Global Agricultural Monitoring (GEOGLAM) initiative, the International Fund for Agricultural Development (IFAD), the International Food Policy Research Institute (IFPRI), the International Grains Council (IGC), the Organisation for Economic Co-operation and Development (OECD), the World Food Program (WFP), the World Trade Organization (WTO), the United Nations Conference on Trade and Development (UNCTAD), the United Nations High-Level Task Force on the Global Food Security Crisis (UN-HLTF), and the World Bank.

AMIS was created as a tool to address excessive food price volatility and to strengthen global food security in a period of heightened insecurity in international food markets. Its creation is thus intrinsically linked to the two consecutive price hikes that occurred in 2007/08 and 2010.

After the 2007-08 world food price crisis led to social unrest in a number of countries and drastically worsened the food security situation, the world experienced another food price shock in the summer of 2010 when the Russian Federation announced an export ban on wheat in response to a severe drought and wildfires that threatened much of the country's crop.

Under the auspices of its Intergovernmental Groups on Grains and Rice, FAO invited all its members to Rome for an extraordinary meeting in September 2010 to discuss the troubled market conditions and to stimulate a coordinated response. While the event failed to yield any immediate results, it can be credited for triggering constructive discussions that eventually led to the creation of AMIS. The meeting acknowledged that unexpected price hikes and volatility were "amongst major threats to food security and that their root causes need to be addressed." In particular it recognized "the lack of reliable and up-to-date information on crop supply and demand and export availability" as well as "insufficient market transparency at all levels including in relation to futures markets" among the main drivers of the most recent disturbances in world food markets. It further emphasized the need "to enhance market information and transparency", calling for improved "monitoring of planting intentions, crop development and domestic market information."

These ideas were taken up during the G20 Summit in Seoul in November 2010, which asked a number of international institutions to identify the best ways to manage and mitigate risks of food price volatility without distorting markets. The ensuing report was presented to the French Presidency of the G20 in June 2011, concluding with a list of ten recommendations, among which to establish AMIS. In the final declaration of the G20 Summit in Cannes, heads of state and government of the G20 countries stressed the importance of improving "market information and transparency in order to make international markets for agricultural commodities more effective." In order to address these challenges, they decided to launch AMIS that was officially inaugurated in September 2011.

Participating Countries

Participants in AMIS include G20 countries plus Spain and seven additional major exporting and importing countries of the AMIS crops. These are: Egypt, Kazakhstan, Nigeria, the Philippines, Thailand, Ukraine, and Vietnam. G20 members are Argentina, Australia, Brazil, Canada, China, European Union, France, Germany, India, Indonesia, Italy, Korea, Japan, Mexico, Russian Federation, Saudi Arabia, South Africa, Turkey, United Kingdom, and the United States.

Objectives

According to the Terms of Reference that established AMIS, the following objectives are central:

- Improve agricultural market information, analysis and short-term supply and demand forecasts at both national and international levels.
- Collect and analyze policy information affecting global commodity markets, and promote international policy dialogue and coordination.
- Report on critical conditions of international food markets, including structural weaknesses, and strengthen global early warning capacity on these movements.

- Build data collection capacity in participating countries by promoting best practices and improved methodologies, providing training to national stakeholders and facilitating the exchange of lessons learned among participating countries.

Structure

AMIS consists of three main bodies:

- The Global Food Market Information Group provides and assesses market and policy information. It unites technical representatives from participating countries who meet twice per year.
- The Rapid Response Forum promotes early discussion among decision-level officials about critical conditions in international food markets, and encourages the coordination of policies. It is composed of senior officials from participating countries who meet once per year, as well as when the market situation warrants international policy action.
- The Secretariat produces short-term market outlooks, assessments and analyses, and supports all functions of the Forum and the Information Group. It is governed by a steering committee that includes one representative from each of the ten member organizations, namely: FAO, IFAD, IFPRI, IGC, OECD, WFP, WTO, UNCTAD, UN-HLTF, and the World Bank.

Data Mining in Agriculture

Data mining in agriculture consists in the application of data mining techniques to agriculture. Recent technologies are nowadays able to provide a lot of information on agricultural-related activities, which can then be analyzed in order to find important information. A related, but not equivalent term is precision agriculture.

Applications

Relationship between Sprays and Fruit Defects

Fruit defects are often recorded (for a multitude of reasons, sometimes for insurance reasons when exporting fruit overseas). It may be done manually or through computer vision (detecting surface defects when grading fruit). Spray diaries are a legal requirement in many countries and at the very least record the date of spray and the product name. It is known that spraying can have affect different fruit defects for different fruit. Fungicidal sprays are often used to prevent rots from being expressed on fruit. It is also known that some sprays can cause russeting on apples. Currently much of this knowledge comes anecdotally, however some efforts have been in regards to the use of data mining in horticulture.

Prediction of Problematic Wine Fermentations

Wine is widely produced all around the world. The fermentation process of the wine is very important, because it can impact the productivity of wine-related industries and also the quality of wine. If the fermentation could be categorized and predicted at the early stages of the process, it could be altered in order to guarantee a regular and smooth fermentation. Fermentations are nowadays studied by using different techniques, such as, for example, the k-means algorithm, and a technique for classification based on the concept of biclustering. Note that these works are different from the ones where a classification of different kinds of wine is performed.

Predicting Metabolizable Energy of Poultry Feed

A group method of data handling-type neural network (GMDH-type network) with an evolutionary method of genetic algorithm was used to predict the metabolizable energy of feather meal and poultry offal meal based on their protein, fat, and ash content. Published data samples were collected from literature and used to train a GMDH-type network model. The novel modeling of GMDH-type network with an evolutionary method of genetic algorithm can be used to predict the metabolizable energy of poltry feed samples based on their chemical content. It is also reported that the GMDH-type network may be used to accurately estimate the poultry performance from their dietary nutrients such as dietary metabolizable energy, protein and amino acids.

Detection of Diseases from Sounds Issued by Animals

The detection of animal's diseases in farms can impact positively the productivity of the farm, because sick animals can cause contaminations. Moreover, the early detection of the diseases can allow the farmer to cure the animal as soon as the disease appears. Sounds issued by pigs can be analyzed for the detection of diseases. In particular, their coughs can be studied, because they indicate their sickness. A computational system is under development which is able to monitor pig sounds by microphones installed in the farm, and which is also able to discriminate among the different sounds that can be detected.

Growth of Sheep from Genes Polymorphism

Polymerase chain reaction-single strand conformation polymorphism (PCR-SSCP) method was used to determine the growth hormone (GH), leptin, calpain, and calpastatin polymorphism in Iranian Baluchi male sheep. An artificial neural network (ANN) model was developed to describe average daily gain (ADG) in lambs from input parameters of GH, leptin, calpain, and calpastatin polymorphism, birth weight, and birth type. The results revealed that the ANN-model is an appropriate tool to recognize the patterns of data to predict lamb growth in terms of ADG given specific genes polymorphism, birth weight, and birth type. The platform of PCR-SSCP approach and ANN-based model analyses may be used in molecular marker-assisted selection and breeding programs to design a scheme in enhancing the efficacy of sheep production.

Sorting Apples by Watercores

Before going to market, apples are checked and the ones showing some defects are removed. However, there are also invisible defects, that can spoil the apple flavor and look. An example of invisible defect is the watercore. This is an internal apple disorder that can affect the longevity of the fruit. Apples with slight or mild watercores are sweeter, but apples with moderate to severe degree of watercore cannot be stored for any length of time. Moreover, a few fruits with severe watercore could spoil a whole batch of apples. For this reason, a computational system is under study which takes X-ray photographs of the fruit while they run on conveyor belts, and which is also able to analyse (by data mining techniques) the taken pictures and estimate the probability that the fruit contains watercores.

Optimizing Pesticide use by Data Mining

Recent studies by agriculture researchers in Pakistan (one of the top four cotton producers of the world) showed that attempts of cotton crop yield maximization through pro-pesticide state policies have led to a dangerously high pesticide use. These studies have reported a negative correlation between pesticide use and crop yield in Pakistan. Hence excessive use (or abuse) of pesticides is harming the farmers with adverse financial, environmental and social impacts. By data mining the cotton Pest Scouting data along with the meteorological recordings it was shown that how pesticide use can be optimized (reduced). Clustering of data revealed interesting patterns of farmer practices along with pesticide use dynamics and hence help identify the reasons for this pesticide abuse.

Pesticide Abuse by Data Mining

To monitor cotton growth, different government departments and agencies in Pakistan have been recording pest scouting, agriculture and metrological data for decades. Coarse estimates of just the cotton pest scouting data recorded stands at around 1.5 million records, and growing. The primary agro-met data recorded has never been digitized, integrated or standardized to give a complete picture, and hence cannot support decision making, thus requiring an Agriculture Data Warehouse. Creating a novel Pilot Agriculture Extension Data Warehouse followed by analysis through querying and data mining some interesting discoveries were made, such as pesticides sprayed at the wrong time, wrong pesticides used for the right reasons and temporal relationship between pesticide usage and day of the week.

Analyzing Chicken Performance Data by Neural Network Models

A platform of artificial neural network-based models with sensitivity analysis and optimization algorithms was used successfully to integrate published data on the responses of broiler chickens to threonine. Analyses of the artificial neural network models for weight gain and feed efficiency from a compiled data set suggested that the dietary protein concentration was more important than the threonine concentration. The results revealed that a diet containing 18.69% protein and 0.73% threonine may lead to producing optimal weight gain, whereas the optimal feed efficiency may be achieved with a diet containing 18.71% protein and 0.75% threonine.

Satellite Crop Monitoring

Satellite crop monitoring is the technology which facilitates real-time crop vegetation index monitoring via spectral analysis of high resolution satellite images for different fields and crops which enables to track positive and negative dynamics of crop development. The difference in vegetation index informs about single-crop development disproportions that speaks for the necessity of additional agriculture works on particular field zones—that is because satellite crop monitoring belongs to precision agriculture methods.

Satellite crop monitoring technology allows to perform online crop monitoring on different fields, located in different areas, regions, even countries and on different continents. The technology's advantage is a high automation level of sown area condition and its interpretation in an interactive map which can be read by different groups of users.

Satellite crop monitoring technology users are:

- Agronomists and agriculture companies management (crop vegetation control, crop yield forecasting, management decisions optimization);
- Business owners (business prospects estimates, making reasonable decisions on capital investments, providing information for management decisions);
- Investors and investment analysts (investment potential estimation, making investment decisions, making sustainable forecasts);
- Insurance brokers (data collection, clients claims verification, scale of rates and insurance premium amounts calculation);
- Agriculture machinery producers (integration of crop monitoring solutions with agriculture machinery board computers operations, functional development);
- State and sectoral organisations engaged in agriculture, food security and ecological problems.

The technology based on spectral analysis of high resolution satellite crop images which enables to monitor vegetation developments, soil temperature, humidity and to reveal problem areas on the field. Satellite crop monitoring is also suitable to precise weather forecast based on concrete field coordinates and to recall historical weather data retrospection.

Along with the development of remote sensing applications, satellite monitoring data has become the uppermost data source to monitor large-scale crop condition based on vegetation index analysis. Vegetation images show crop growth from planting through to harvest, changes as the season progresses and abnormalities such as weed patches, soil compaction, watering problems etc. A georeferenced and orthorectified image can locate these problem areas as well as the size of the area affected can be easily determined. Satellite crop monitoring and vegetation control help the farmer make informed decisions about the most feasible solution. In addition to highlighting problematic areas, images will also help monitor the effectiveness of any corrective actions which may be implemented. Images can act as an early indicator of crop yield. This early predictor of yield can aid the farmer in making marketing decisions as well as the allocation of resources.

To gain the benefits from satellite crop monitoring data farmers, managers, consultants and technicians must understand and be able to interpret the image. There are a wide range of enhancement tools available which can help make an image more interpretable for specific applications. Enhancement and classification tools are often used to highlight features. The techniques employed will depend on the type of remote sensed data as well as the objectives of the end user.

The satellite provides imagery data at different spatial, spectral and temporal resolutions for agriculture and crop assessment, crop health, change detection, environmental analysis, irrigated landscape mapping, yield determination and soils analysis. Scheduling and timing of image acquisition is very important and will hinge on the main goals and the type of information that the end user is hoping to gain. Images can show variations in organic matter and drainage patterns. Soils higher in organic matter can be differentiated from lighter sandier soil that has a lower organic matter content. This geospatial information is valuable when used in conjunction with ancillary data to define management zones for a field.

Agricultural Machinery

Agricultural machinery is machinery used in farming or other agriculture. There are many types of such equipment, from hand tools and power tools to tractors and the countless kinds of farm implements that they tow or operate. Diverse arrays of equipment are used in both organic and nonorganic farming. Especially since the advent of mechanised agriculture, agricultural machinery is an indispensable part of how the world is fed.

A German combine harvester by Claas.

A British crop sprayer by Lite-Trac.

The Industrial Revolution

With the coming of the Industrial Revolution and the development of more complicated machines, farming methods took a great leap forward. Instead of harvesting grain by hand with a sharp blade, wheeled machines cut a continuous swath. Instead of threshing the grain by beating it with sticks, threshing machines separated the seeds from the heads and stalks. The first tractors appeared in the late 19th century.

Steam Power

Power for agricultural machinery was originally supplied by ox or other domesticated animals. With the invention of steam power came the portable engine, and later the traction engine, a multipurpose, mobile energy source that was the ground-crawling cousin to the steam locomotive. Agricultural steam engines took over the heavy pulling work of oxen, and were also equipped with a pulley that could power stationary machines via the use of a long belt. The steam-powered machines were low-powered by today's standards but, because of their size and their low gear ratios, they could provide a large drawbar pull. Their slow speed led farmers to comment that tractors had two speeds: "slow, and damn slow."

Internal Combustion Engines

The internal combustion engine; first the petrol engine, and later diesel engines; became the main source of power for the next generation of tractors. These engines also contributed to the development of the self-propelled, combined harvester and thresher, or combine harvester (also shortened to 'combine'). Instead of cutting the grain stalks and transporting them to a stationary threshing machine, these combines cut, threshed, and separated the grain while moving continuously through the field.

Types

Combines might have taken the harvesting job away from tractors, but tractors still do the majority of work on a modern farm. They are used to push/pull implements—machines that till the ground, plant seed, and perform other tasks.

A John Deere cotton harvester at work in a cotton field.

Tillage implements prepare the soil for planting by loosening the soil and killing weeds or competing plants. The best-known is the plow, the ancient implement that was upgraded in 1838 by John Deere. Plows are now used less frequently in the U.S. than formerly, with offset disks used instead to turn over the soil, and chisels used to gain the depth needed to retain moisture.

From left to right: John Deere 7800 tractor with Houle slurry trailer, Case IH combine harvester, New Holland FX 25 forage harvester with corn head.

The most common type of seeder is called a planter, and spaces seeds out equally in long rows, which are usually two to three feet apart. Some crops are planted by drills, which put out much more seed in rows less than a foot apart, blanketing the field with crops. Transplanters automate the task of transplanting seedlings to the field. With the widespread use of plastic mulch, plastic mulch layers, transplanters, and seeders lay down long rows of plastic, and plant through them automatically.

A New Holland TR85 combine harvester.

After planting, other implements can be used to cultivate weeds from between rows, or to spread fertilizer and pesticides. Hay balers can be used to tightly package grass or alfalfa into a storable form for the winter months.

Modern irrigation relies on machinery. Engines, pumps and other specialized gear provide water quickly and in high volumes to large areas of land. Similar types of equipment can be used to deliver fertilizers and pesticides.

Besides the tractor, other vehicles have been adapted for use in farming, including trucks, airplanes, and helicopters, such as for transporting crops and making equipment mobile, to aerial spraying and livestock herd management.

New Technology and the Future

The basic technology of agricultural machines has changed little in the last century. Though modern harvesters and planters may do a better job or be slightly tweaked from their predecessors, the US$250,000 combine of today still cuts, threshes, and separates grain in the same way it has always been done. However, technology is changing the way that humans operate the machines, as computer monitoring systems, GPS locators and self-steer programs allow the most advanced tractors and implements to be more precise and less wasteful in the use of fuel, seed, or fertilizer. In the foreseeable future, there may be mass production of driverless tractors, which use GPS maps and electronic sensors.

Open Source Agricultural Equipment

Many farmers are upset by their inability to fix the new types of high-tech farm equipment. This is due mostly to companies using intellectual property law to prevent farmers from having the legal right to fix their equipment (or gain access to the information to allow them to do it). In October 2015 an exemption was added to the DMCA to allow inspection and modification of the software in cars and other vehicles including agricultural machinery.

The Open Source Agriculture movement counts different initiatives and organizations such as Farm Labs which is a network in Europe, l'Atelier Paysan which is a cooperative to teach farmers in France how to build and repair their tools, and Ekylibre which is an open-source company to provide farmers in France with open source software (SaaS) to manage farming operations. In the United States, the MIT Media Lab's Open Agriculture Initiative seeks to foster "the creation of an open-source ecosystem of technologies that enable and promote transparency, networked experimentation, education, and hyper-local production". It develops the Personal Food Computer, an educational project to create a "controlled environment agriculture technology platform that uses robotic systems to control and monitor climate, energy, and plant growth inside of a specialized growing chamber". It includes the development of Open Phenom, an open source library with open data sets for climate recipes which link the phenotype response of plants (taste, nutrition) to environmental variables, biological, genetic and resource-related necessary for cultivation (input). Plants with the same genetics can naturally vary in color, size, texture growth rate, yield, flavor and nutrient density according to the environmental conditions in which they are produced.

Agricultural Drone

A UAV (Unmanned Aerial Vehicle) is a flying device that can fly a pre-set course with the help of an autopilot and GPS coordinates. The device also has normal radio controls; it can be piloted manually in case of a fault or dangerous situation. Sometimes the term UAV is used to refer to the complete system, including ground stations and video systems, however the term is most commonly used for model planes and helicopters with both fixed and rotary wings.

Advantages

Unmanned Aerial Vehicle offers less stressful environment, it is used for better decision making, it presents safer environment, and they can fly longer hours as long as the vehicle allows for it (no human fatigue in the plane).

There is no need for the qualified pilot to fly it, in the long run. Unmanned Air Vehicle can stay in the air for up to 30 hours, doing the repetitive tasks, performing the precise, repetitive raster scan of the region, day-after-day, night-after-night in the complete darkness or in the fog and under computer control.

Unmanned Air Vehicle performs the geological survey, it performs the visual or thermal imaging of the region, it can measure the cell phone, radio or TV coverage over any terrain, the drone pilots or operators can easily hand off controls of the drone without any operational downtime.

The drones can have more pinpoint accuracy from greater distances.

Basic Principle

The 4 propellers of a drone or quadcopter are fixed and vertically orientated. Each propeller has a variable and independent speed which allows a full range of movements. The core components of a drone are as follows:

Chassis: The skeleton of the drone which all componentry is fixed to. The chassis design is a trade-off between strength (especially when additional weights such as cameras are attached) and additional weight, which will require longer propellers and stronger motors to lift.

Propellers: Principally effect load the drone can carry the speed it can fly and the speed it can manoeuvre. The length can be modified; longer propellers can achieve greater lift at a lower rpm but take longer to speed up/slow down. Shorter propellers can change speed quicker and thus are more manoeuvrable; however they require a higher rotational speed to achieve the same power as longer blades. This causes excess motor strain and thus reduces motor life span. A more aggressive pitch will allow quicker movement but reduced hovering efficiency.

Motors: 1 per propeller, drone motors are rated in “kV” units which equates to the number of revolutions per minute it can achieve when a voltage of 1 volt is supplied to the motor with no load. A faster motor spin will give more flight power, but requires more power from the battery resulting in a decreased flight time.

Electronic Speed Controller (ESC): provides a controlled current to each motor to produce the correct spin speed and direction.

Flight Controller: The onboard computer which interprets incoming signals sent from the pilot and sends corresponding inputs to the ESC to control the quadcopter.

Radio Receiver: Receives the control signals from the pilot.

Battery: Generally lithium polymer batteries are used due to high power density and ability to recharge.

Further to this, sensors can be used such as accelerometers, gyroscopes, GPS and barometers for positional measurements. Cameras are also frequently mounted for navigation and aerial photography.

Drone Mechanism

A drone is controlled manually with a hand- held radio control transmitter which manually controls the propellers. Sticks on the controller allow movements in different directions and trim buttons allow the trim to be adjusted to balance the drone. Screens can also be used to receive live video footage from the on-board camera and to display sensor data.

Further to this, on-board sensors can provide helpful settings such as:

- Auto altitude where the drone will move at a fixed altitude, and;
- GPS hold, where the drone will remain at a fixed GPS position.

Drone can also be flown autonomously, modern flight controllers can use software to mark GPS waypoints that the vehicle will fly to and land or move to a set altitude.

Agricultural Applications of Drone

Soil and field analysis: Drones can be instrumental at the start of the crop cycle. They produce precise 3-D maps for early soil analysis, useful in planning seed planting patterns. After planting, drone-driven soil analysis provides data for irrigation and nitrogen-level management.

Planting: Startups have created drone-planting systems that achieve an uptake rate of 75 percent and decrease planting costs by 85 percent. These systems shoot pods with seeds and plant nutrients into the soil, providing the plant all the nutrients necessary to sustain life.

Crop spraying: Drones can scan the ground and spray the correct amount of liquid, modulating distance from the ground and spraying in real time for even coverage. The result: increased efficiency with a reduction of in the amount of chemicals penetrating into groundwater. In fact, experts estimate that aerial spraying can be completed up to five times faster with drones than with traditional machinery.

Crop monitoring: Vast fields and low efficiency in crop monitoring together create farming's largest obstacle. Monitoring challenges are exacerbated by increasingly unpredictable weather conditions, which drive risk and field maintenance costs.

Irrigation: Drones with hyper-spectral, multispectral, or thermal sensors can identify which parts of a field are dry or need improvements. Additionally, once the crop is growing, drones allow the calculation of the vegetation index, which describes the relative density and health of the crop, and show the heat signature, the amount of energy or heat the crop emits.

Health assessment: It's essential to assess crop health and spot bacterial or fungal infections on trees. By scanning a crop using both visible and near-infrared light, drone-carried devices can identify which plants reflect different amounts of green light and

NIR light. This information can produce multispectral images that track changes in plants and indicate their health.

At beginning of 21st century, people looked forward to new millennium. No one could imagine what kind of new technology take place. In that way, Agricultural drone is an amazing advance technology, which is becoming a tool like any agricultural equipment. Various reasons have behind this, like comparatively cheap agricultural drone with advanced imaging capabilities and sensor are giving specific data to the farmer. By using this data, farmer can increase yields and reduce crop damage. Moreover, less use of pesticides reduces environmental damages.

However, Farming is an input-output problem. With using of drone, farmer can reduce inputs – water and pesticides and maintaining same output, it will be overcoming the food shortage. Agricultural drone changes farmer's ability to monitor and manage the key aspect of farm business that is impossible to sustain in remote place. Conclusively, we can say Drone, which started as a military technology may end up better known as a green-tech technology.

Irrigation Systems

Surface Irrigation

Surface irrigation is the application of water by gravity flow to the surface of the field. Either the entire field is flooded (basin irrigation) or the water is fed into small channels (furrows) or strips of land (borders).

Worldwide there are approx. 3.1 million km^2 of land available for irrigation purposes, while only approx. 2.6 million km^2 are utilized. With about 95% share of the total irrigation worldwide, surface irrigation is by far the most widespread irrigation method. It is normally used when conditions are favorable: mild and regular slopes, soil type with medium to low infiltration rate, and a sufficient supply of surface or groundwater.

Surface irrigation is widely utilized and therefore a well-known system which can be operated without any high-tech applications. In general, it is more labor intensive than other irrigation methods. Proper design of surface irrigation systems takes into account the soil type (texture and infiltration rate), slope, and levelness of the field, stream size, and length of run. Leveling the fields and building the water ditches and reservoirs might be expensive, but once this is done, costs are low and the self-help capacity is very high.

Basin Irrigation

Basins are flat areas of land, surrounded by low bunds. The bunds prevent the water from flowing to the adjacent fields. Basin irrigation is commonly used for rice (paddy) grown on flat lands or in terraces on hillsides. Trees (e.g. citrus, banana) can also be grown in basins, where one tree is usually located in the middle of a small basin.

Furrow Irrigation

Furrows are small channels, which carry water down the land slope between the crop rows. Water infiltrates into the soil as it moves along the slope. The crop is usually grown on the ridges between the furrows. This method is suitable for all row crops and for crops that cannot stand in water for long periods (e.g. 12-24 hours).

To manage the irrigation with the traditional breach system, the farmer has to open and close the embankment of the water-conveying channel. This is the most common method of releasing water from a channel, but it can also be the most damaging. Not only is it difficult to control the discharge, but there can be serious erosion of the channel embankment. If other more controllable methods are available, then these should be used in preference to this.

Border Irrigation

Borders are long, sloping strips of land separated by bunds. They are sometimes called border strips.

Irrigation water can be fed to the border in several ways: opening up the channel bank using small outlets or gates or by means of siphons or spiles. Siphons are small curved pipes that deliver water over a ditch bank. Spiles are small pipes buried in the ditch bank. A sheet of water flows down the slope of the border, guided by the bunds on either side. When the desired amount of water has been delivered to the border, the stream is turned off. However, if the flow is stopped too soon, there may not be enough water in the border to complete the irrigation at the far end. If it is left running for too long, the water may run off the end of the border and be lost in the drainage system.

Border irrigation is best suited to the larger mechanized farms as it is designed to produce long uninterrupted field lengths for ease of machine operations. Borders can be up to 800 m or more in length and 3 - 30 m wide depending on a variety of factors.

Border slopes should be uniform, with a minimum slope of 0.05% to provide adequate drainage and a maximum slope of 2% to limit problems of soil erosion. Deep homogenous loam or clay soils with medium infiltration rates are preferred. On heavy clay soils, border irrigation may cause problems because of the low infiltration rates (basin irrigation is more suited on these soil types). Close growing crops such as pasture or alfalfa are preferred.

Advantages and Disadvantages

Level surface irrigation (water is ponded on an enclosed level field and allowed to infiltrate in basins, borders, or furrows) has some distinct advantages which can be summarized as follows:

- Irrigation management is very easy and does not require modern technology and can largely build on local traditional knowledge;
- Adapts well to small land holdings and does not require high financial input;
- Adapts easily to flat topography and can function without outlet drainage facilities;
- Works well with short-term water supplies;
- Irrigation allows full utilization of rainwater and can achieve high application efficiencies;
- Adapts well to moderate to low infiltration rates and allows easy leaching of salts.

Disadvantages of level surface irrigation include:

- Requires level land to achieve high efficiencies (maximum land elevation fluctuation should not be greater than half the applied irrigation depth);
- Soils with high infiltration rates require small field sizes, which interferes with mechanization;
- Difficulty to apply small irrigation quantities, excess water is difficult to evacuate, particularly during times of excess rainfall;
- Plants are partly covered with water sometimes over extended periods (in low infiltration rate soils);

- Small basins require extensive delivery channels and are not easily adaptable to tractor mechanization.

Graded surface irrigation (water is fed into the high end of a field and is allowed to run slowly to the low end) shows the following advantages:

- Requires low capital and energy costs;
- Allows irrigation on sloping land (as is found in many irrigated areas);
- Allows irrigation of long fields with relatively small flows;
- Is applicable to soils with moderate to fairly high infiltration rates;
- Field drainage of excess water is made possible.

Disadvantages of graded surface irrigation include:

- A high degree of management and water control is required to achieve high irrigation efficiencies;
- High irrigation efficiencies require uniformly graded and shaped land;
- With moderate to slow infiltration rates, long irrigation times are required;
- Except for soils with high infiltration rates, a drainage outlet must be available from every field to dispose of tail water and rainwater;
- Labor intensive.

A special and rarely use approach is the subsurface irrigation - Water is directed to the subsoil (crop root zone) in the area to be irrigated. The water is used to artificially control the groundwater table, and is normally delivered through perforated pipes buried in the ground. In some Latin American countries, porous clay pots are buried in the ground and filled with water that slowly seeps into the subsoil to moisten the roots of crops.

Localized Irrigation

Localized irrigation systems apply water directly where the plant is growing thus minimizing water loss through evaporation from the soil. Such localized irrigation systems include drip irrigation, porous clay pots, porous pipes, and perforated plastic sleeves.

Porous Clay Pots

This is a method of irrigation in which water is stored in clay pots buried in the ground, from where it is slowly released to the plants. This method is good for fruit trees. Such use of soil-embedded porous jars is one of the oldest continuous irrigation methods that probably originated in the Far East and North Africa.

The method consists of:

- Clay pots that are placed in shallow pits dug for this purpose;
- Soil is then packed around the neck of the pots so that the necks protrude a few centimetres above the ground surface;
- Water is poured into the pots, either by hand or by means of a flexible hose connected to a water source.

The pots are made of locally available clay with optimum properties of strength (to resist crushing), permeability (to exude water into the soil at an approximately steady rate), and size (to hold enough water for at least one day's supply).

The potential of clay-pot irrigation has not been fully exploited by farmers in the eastern and southern Africa region, even though the technology is suitable for small scale farmers. There have not been many reports of previous experience in the region.

Porous Clay Pipes

Water is spread along a continuous horizontal band in the soil. This method is most suited to closely spaced crops such as vegetables. The locally made clay pipes are approximately 24 cm in length and 7.5 cm in internal diameter, with wall thickness of 2 cm. The pipes are placed at the bottom of a shallow trench (about 25 cm deep) representing the centre line of a 1-m wide bed.

Perforated Plastic Sleeves

Plastic sheeting has been used to make a sleeve-like casing. The advantage of this is the low cost. However, the method has several distinct disadvantages that restrict the range of its applicability. Since the soft plastic material used for making the sleeve does not retain its shape, the sleeve must be filled with sand before being placed in the soil. The sand filling reduces the capacity of the sleeve by some 50–60%. Moreover, the sand itself tends to retain a significant fraction of the moisture and thus to restrict outflow.

Since the plastic casing is essentially impervious, it must be perforated. The difficulty of standardizing the diameter and density of the perforations introduces another variable unto the system. The best configuration must be established by trial and error. This method has been used with success in Senegal. An interesting variation in Sri Lanka consists of a 50-cm long PVC pipe of standard °-inch diameter. The pipe ends in an emitter, a block made from a 1:10 cement/sand mixture. The design was found to be effective in enhancing the survival and growth rate of young fruit trees.

Drip Irrigation

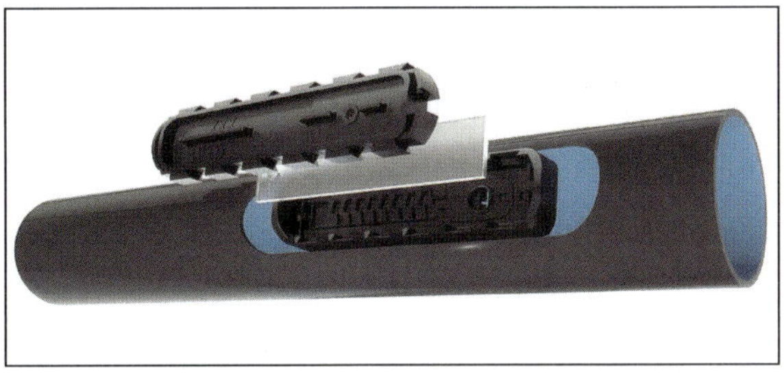

Open pressure compensated dripper.

Drip irrigation is a type of micro-irrigation system that has the potential to save water and nutrients by allowing water to drip slowly to the roots of plants, either from above the soil surface or buried below the surface. The goal is to place water directly into the root zone and minimize evaporation. Drip irrigation systems distribute water through a network of valves, pipes, tubing, and emitters. Depending on how well designed, installed, maintained, and operated it is, a drip irrigation system can be more efficient than other types of irrigation systems, such as surface irrigation or sprinkler irrigation.

Current Developments

Careful study of all the relevant factors like land topography, soil, water, crop and agro-climatic conditions are needed to determine the most suitable drip irrigation system and components to be used in a specific installation.

Micro-spray Heads

Drip irrigation may also use devices called micro-spray heads, which spray water in a small area, instead of dripping emitters. These are generally used on tree and vine crops with wider root zones.

Subsurface Drip Irrigation

Subsurface drip irrigation (SDI) uses permanently or temporarily buried dripperline or drip tape located at or below the plant roots. It is becoming popular for row crop irrigation, especially in areas where water supplies are limited, or recycled water is used for irrigation.

Components and Operation

Components used in drip irrigation (listed in order from water source) include:

- Pump or pressurized water source;
- Water filter(s) or filtration systems: Sand separator, Fertigation systems (Venturi injector) and chemigation equipment (optional);

- Backwash controller (Backflow prevention device);
- Pressure Control Valve (pressure regulator);
- Distribution lines (main larger diameter pipe, maybe secondary smaller, pipe fittings);
- Hand-operated, electronic, or hydraulic control valves and safety valves;
- Smaller diameter polyethylene tube (often called "laterals");
- Poly fittings and accessories (to make connections);
- Emitting devices at plants (emitter or dripper, micro spray head, inline dripper or inline drip tube).

Nursery flowers watered with drip irrigation in Israel.

Horticulture drip emitter in a pot.

In drip irrigation systems, pump and valves may be manually or automatically operated by a controller.

Most large drip irrigation systems employ some type of filter to prevent clogging of the small emitter flow path by small waterborne particles. New technologies are now being offered that minimize clogging. Some residential systems are installed without additional filters since potable water is already filtered at the water treatment plant. Virtually all drip irrigation equipment manufacturers recommend that filters be employed and generally will not honor warranties unless this is done. Last line filters just before the final delivery pipe are strongly recommended in addition to any other filtration system due to fine particle settlement and accidental insertion of particles in the intermediate lines.

Drip and subsurface drip irrigation is used almost exclusively when using recycled municipal wastewater. Regulations typically do not permit spraying water through the air that has not been fully treated to potable water standards.

Because of the way the water is applied in a drip system, traditional surface applications of timed-release fertilizer are sometimes ineffective, so drip systems often mix liquid fertilizer with the irrigation water. This is called fertigation; fertigation and chemigation (application of pesticides and other chemicals to periodically clean out the system, such as chlorine or sulfuric acid) use chemical injectors such as diaphragm pumps, piston pumps, or aspirators. The chemicals may be added constantly whenever the system is irrigating or at intervals. Fertilizer savings of up to 95% are being reported from recent university field tests using drip fertigation and slow water delivery as compared to timed-release and irrigation by micro spray heads.

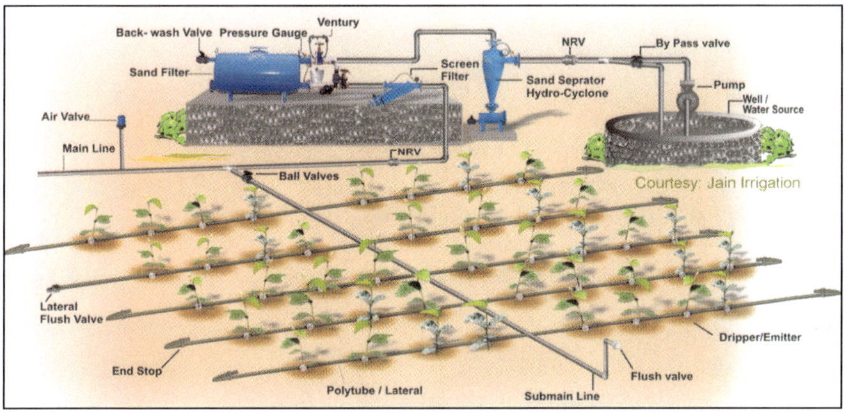

Drip irrigation system layout and its parts.

Properly designed, installed, and managed, drip irrigation may help achieve water conservation by reducing evaporation and deep drainage when compared to other types of irrigation such as flood or overhead sprinklers since water can be more precisely applied to the plant roots. In addition, drip can eliminate many diseases that are spread through water contact with the foliage. Finally, in regions where water supplies are severely limited, there may be no actual water savings, but rather simply an increase in production while using the same amount of water as before. In very arid regions or on sandy soils, the preferred method is to apply the irrigation water as slowly as possible.

Pulsed irrigation is sometimes used to decrease the amount of water delivered to the plant at any one time, thus reducing runoff or deep percolation. Pulsed systems are typically expensive and require extensive maintenance. Therefore, the latest efforts by emitter manufacturers are focused on developing new technologies that deliver irrigation water at ultra-low flow rates, i.e. less than 1.0 liter per hour. Slow and even delivery further improves water use efficiency without incurring the expense and complexity of pulsed delivery equipment.

An emitting pipe is a type of drip irrigation tubing with emitters pre-installed at the factory with specific distance and flow per hour as per crop distance.

An emitter restricts water flow passage through it, thus creating head loss required (to the extent of atmospheric pressure) in order to emit water in the form of droplets. This head loss is achieved by friction/turbulence within the emitter.

Advantages and Disadvantages

The advantages of drip irrigation are:

- Fertilizer and nutrient loss is minimized due to a localized application and reduced leaching.
- Water application efficiency is high if managed correctly.
- Field leveling is not necessary.
- Fields with irregular shapes are easily accommodated.
- Recycled non-potable water can be safely used.
- Moisture within the root zone can be maintained at field capacity.
- Soil type plays a less important role in the frequency of irrigation.
- Soil erosion is lessened.
- Weed growth is lessened.
- Water distribution is highly uniform, controlled by the output of each nozzle.
- Labour cost is less than other irrigation methods.
- Variation in supply can be regulated by regulating the valves and drippers.
- Fertigation can easily be included with minimal waste of fertilizers.
- Foliage remains dry, reducing the risk of disease.
- Usually operated at lower pressure than other types of pressurized irrigation, reducing energy costs.

The disadvantages of drip irrigation are:

- Initial cost can be more than overhead systems.
- The sun can affect the tubes used for drip irrigation, shortening their lifespan.
- The risks of degrading plastic affecting the soil content and food crops. With many types of plastic, when the sun degrades the plastic, causing it to become brittle, the estrogenic chemicals (that is, chemicals replicating female hormones) which would cause the plastic to retain flexibility have been released into the surrounding environment.
- If the water is not properly filtered and the equipment not properly maintained, it can result in clogging or bioclogging.

- For subsurface drip the irrigator cannot see the water that is applied. This may lead to the farmer either applying too much water (low efficiency) or an insufficient amount of water, this is particularly common for those with less experience with drip irrigation.
- Drip irrigation might be unsatisfactory if herbicides or top dressed fertilizers need sprinkler irrigation for activation.
- Drip tape causes extra cleanup costs after harvest. Users need to plan for drip tape winding, disposal, recycling or reuse.
- Waste of water, time and harvest, if not installed properly. These systems require careful study of all the relevant factors like land topography, soil, water, crop and agro-climatic conditions, and suitability of drip irrigation system and its components.
- In lighter soils subsurface drip may be unable to wet the soil surface for germination. Requires careful consideration of the installation depth.
- Most drip systems are designed for high efficiency, meaning little or no leaching fraction. Without sufficient leaching, salts applied with the irrigation water may build up in the root zone, usually at the edge of the wetting pattern. On the other hand, drip irrigation avoids the high capillary potential of traditional surface-applied irrigation, which can draw salt deposits up from deposits below.
- The PVC pipes often suffer from rodent damage, requiring replacement of the entire tube and increasing expenses.
- Drip irrigation systems cannot be used for damage control by night frosts (like in the case of sprinkler irrigation systems).

Drip Tape

Drip tape is a type of thin-walled dripperline used in drip irrigation. The first drip tape was known as "Dew Hose".

Drip tape duct tape is made of polyethylene and is sold flat on reels. The wall thickness typically ranges from 4 to 25 mils (0.1–0.6 mm). Thicker walled tapes are commonly used for permanent subsurface drip irrigation and thinner walled tapes for temporary throw-away type systems in high-value crops.

Water exits from tape through emitters or drippers. The typical emitter spacing ranges from 6 to 24 inches (150–600 mm). In some products, the emitters are manufactured simultaneously with the tape and are actually formed as part of the product itself. In others, the emitters are manufactured separately and installed at the time of production.

Some product is not a tape, but a thin-walled dripperline, but in popular parlance, both types of products are called tapes. Typical tape diameters are 5/8", 7/8", and 1-3/8", with the larger diameters more commonly used on permanent installations with longer runs.

Drip tape is a recyclable material and can be recycled into viable plastic resins for reuse in the plastics manufacturing industry.

Uses

Irrigation dripper.

Drip irrigation is used in farms, commercial greenhouses, and residential gardens. Drip irrigation is adopted extensively in areas of acute water scarcity and especially for crops and trees such as coconuts, containerized landscape trees, grapes, bananas, ber, egg-plant, citrus, strawberries, sugarcane, cotton, maize, and tomatoes.

Drip irrigation for garden available in drip kits are increasingly popular for the home-owner and consist of a timer, hose and emitter. Hoses that are 4 mm in diameter are used to irrigate flower pots.

Sprinkler Irrigation

In the sprinkler method of irrigation, water is sprayed into the air and allowed to fall on the ground surface somewhat resembling rainfall. The spray is developed by the flow of water under pressure through small orifices or nozzles. The pressure is usually obtained by pumping. With careful selection of nozzle sizes, operating pressure and sprinkler spacing the amount of irrigation water required to refill the crop root zone can be applied nearly uniform at the rate to suit the infiltration rate of soil.

Advantages of Sprinkler Irrigation

- Elimination of the channels for conveyance, therefore no conveyance loss;
- Suitable to all types of soil except heavy clay;

- Suitable for irrigating crops where the plant population per unit area is very high. It is most suitable for oil seeds and other cereal and vegetable crops;
- Water saving;
- Closer control of water application convenient for giving light and frequent irrigation and higher water application efficiency;
- Increase in yield;
- Mobility of system;
- May also be used for undulating area;
- Saves land as no bunds etc. are required;
- Influences greater conducive micro-climate;
- Areas located at a higher elevation than the source can be irrigated;
- Possibility of using soluble fertilizers and chemicals;
- Less problem of clogging of sprinkler nozzles due to sediment laden water.

Crop Response to Sprinkler

The trials conducted in different parts of the country revealed water saving due to sprinkler system varies from 16 to 70 % over the traditional method with yield increase from 3 to 57 % in different crops and agro climatic conditions.

Response of Different Crops to Sprinkler Irrigation

Crops	Water Saving, %	Yield increase, %
Bajra	56	19
Barley	56	16
Bhindi	28	23
Cabbage	40	3
Cauliflower	35	12
Chillies	33	24
Cotton	36	50
Cowpea	19	3
Fenugreek	29	35

Garlic	28	6
Gram	69	57
Groundnut	20	40
Jowar	55	34
Lucerne	16	27
Maize	41	36
Onion	33	23
Potato	46	4
Sunflower	33	20
Wheat	35	24

General Classification of Different Types of Sprinkler Systems

Sprinkler systems are classified into the following two major types on the basis of the arrangement for spraying irrigation water.

- Rotating head or revolving sprinkler system.
- Perforated pipe system.

Rotating Head

Small size nozzles are placed on riser pipes fixed at uniform intervals along the length of the lateral pipe and the lateral pipes are usually laid on the ground surface. They may also be mounted on posts above the crop height and rotated through 90°, to irrigate a rectangular strip. In rotating type sprinklers, the most common device to rotate the sprinkler heads is with a small hammer activated by the thrust of water striking against a vane connected to it.

Example of a few rotating type sprinkler irrigation systems.

Perforated Pipe System

This method consists of drilled holes or nozzles along their length through which water is sprayed under pressure. This system is usually designed for relatively low pressure

(1 kg/cm^2). The application rate ranges from 1.25 to 5 cm per hour for various pressure and spacing.

Based on the portability, sprinkler systems are classified into the following types:

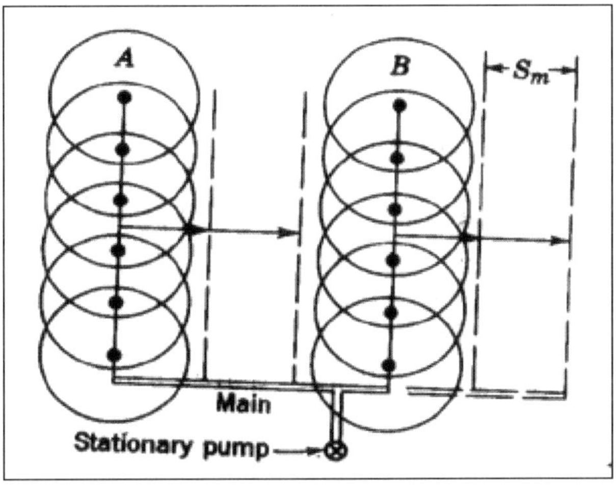

Fully portable sprinkler irrigation system.

- Portable system : A portable system has portable main lines, laterals and pumping plan.
- Semi portable system: A semi portable system is similar to a portable system except that the location of water source and pumping plant is fixed.
- Semi permanent system: A semi permanent system has portable lateral lines, permanent main lines and sub mains and a stationery water source and pumping plant.
- Solid set system: A solid set system has enough laterals to eliminate their movement. The laterals are positions in the field early in the crop season and remain for the season.
- Permanent system: A fully permanent system consists of permanently laid mains, sub mains and laterals and a stationery water source and pumping plant.

Components of Sprinkler Irrigation System

The components of portable sprinkler system are shown through figure.

A sprinkler system usually consists of the following components:

- A pump unit;
- Tubings-main/submains and laterals;
- Couplers;

- Sprinker head;
- Other accessories such as valves, bends, plugs and risers.

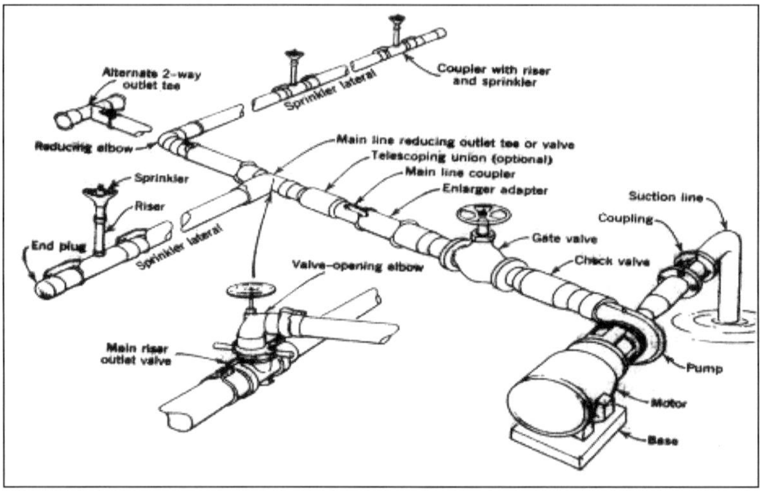

Component of a portable sprinkler irrigation system.

Pumping Unit: Sprinkler irrigation systems distribute water by spraying it over the fields. The water is pumped under pressure to the fields. The pressure forces the water through sprinklers or through perforations or nozzles in pipelines and then forms a spray. A high speed centrifugal or turbine pump can be used for operating sprinkler irrigation for individual fields. Centrifugal pump is used when the distance from the pump inlet to the water surface is less than eight meters. For pumping water from deep wells or more than eight meters, a turbine pump is suggested. The driving unit may be either an electric motor or an internal combustion engine.

Tubings: Mains/submains and laterals: The tubings consist of mainline, submanins and laterals. Main line conveys water from the source and distributes it to the submains. The submains convey water to the laterals which in turn supply water to the sprinklers. Aluminum or PVC pipes are generally used for portable systems, while steel pipes are usually used for center-pivot laterals. Asbestos, cement, PVC and wrapped steel are usually used for buried laterals and main lines.

Couplers: Couplers are used for connecting two pipes and uncoupling quickly and easily. Essentially a coupler should provide:

- A reuse and flexible connection;
- Not leak at the joint;
- Be simple and easy to couple and uncouple;
- Be light, non-corrosive, durable.

Sprinkler Head: Sprinkler head distribute water uniformly over the field without runoff or excessive loss due to deep percolation. Different types of sprinklers are available. They are either rotating or fixed type. The rotating type can be adapted for a wide range of application rates and spacing. They are effective with pressure of about 10 to 70 m head at the sprinkler. Pressures ranging from 16 to 40 m head are considered the most practical for most farmers.

Fixed head sprinklers are commonly used to irrigate small lawns and gardens. Perforated lateral lines are sometimes used as sprinklers. They require less pressure than rotating sprinklers. They release more water per unit area than rotating sprinklers. Hence fixed head sprinklers are adaptable for soils with high intake rate.

Fittings and accessories: The following are some of the important fittings and accessories used in sprinkler system.

- Water meters: It is used to measure the volume of water delivered. This is necessary to operate the system to give the required quantity of water.
- Flange, couplings and nipple used for proper connection to the pump, suction and delivery.
- Pressure gauge: It is necessary to know whether the sprinkler system is working with desired pressure to ensure application uniformity.
- Bend, tees, reducers, elbows, hydrants, butterfly valve and plugs.

Fertilizer applicator: Soluble chemical fertilizers can be injected into the sprinkler system and applied to the crop. The equipment for fertiliser application is relatively cheap and simple and can be fabricated locally. The fertilizer applicator consists of a sealed fertilizer tank with necessary tubings and connections. A venturi injector can be arranged in the main line, which creates the differential pressure suction and allows the fertilizer solution to flow in the main water line.

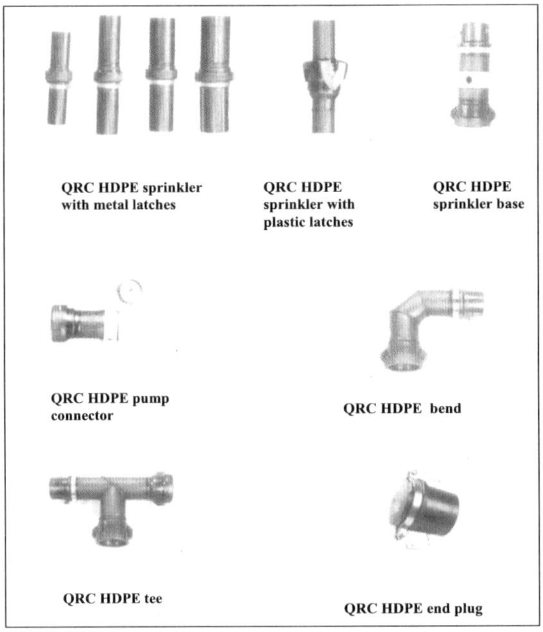

Different sprinkler pipes and fittings.

General Rules for Sprinkler System Design

- Main should be laid up and down hill;
- Lateral should be laid across the slope or nearly on the contour;
- For multiple lateral operation, lateral pipe sizes should not be more than two diameter;
- Water supply source should be nearest to the center of the area;
- Layout should facilitate and minimize lateral movement during the season;
- Booster pump should be considered where small portion of field would require high pressure at the pump;
- Layout should be modified to apply different rates and amounts of water where soils are greatly different in the design area.

While selecting a sprinkler system, the most important physical parameters to be considered are:

- The crop or crops to be cultivated;
- The shape and size (acres) of the field;
- The topography of the field;
- The amount of time and labor required to operate the system.

Selecting Sprinkler System Capacity

A sprinkler system must be designed to apply water uniformly without runoff or erosion. The application rate of the sprinkler system must be matched to the infiltration rate of the most restrictive soil in the field. If the application rate exceeds the soil intake rate, the water will runoff the field or relocate within the field resulting in over and under watered areas.

The sprinkler system capacity is the flow rate needed to adequately irrigate an area and is expressed in liters per minute per acre. The system capacity depends upon on the: Peak crop water requirements during the growing season; effective crop rooting depth; texture and infiltration rate of the soil; the available water holding capacity of the soil; pumping capacity of the well or wells (if wells are the water source).

Constraints in Application of Sprinkler Irrigation

- Uneven water distribution due to high winds;
- Evaporation loss when operating under high temperatures;
- Highly impermeable soils are not suitable;
- Initial cost is high;
- Proper design;
- Lack of Package of practices;
- Lack of awareness;
- Lack of social concern to save natural resources;
- High water pressure required in sprinkler ($>2.5 kg/cm^2$);
- Difficulty in Irrigation During Wind in Sprinkler.

Operation and Maintenance of Sprinkler Systems

Proper design of a sprinkler system does not in itself ensure success. It should be ensured that the prime mover and the pump are in alignment, particularly in the case of tractor-driven pumps. For these the drive shaft as well as the pump shaft should lie at nearly the same height to prevent too great an angle on the universal shaft.

While laying the main and lateral pipes, always begin laying at the pump. This necessarily gives the correct connection of all quick coupling pipes. While joining couplings, it is ensured that both the couplings and the rubber seal rings are clean.

In starting the sprinkler system, the motor or engine is started with the valves closed. The pump must attain the pressure stated on type-plate or otherwise there is a fault in the suction line. After the pump reaches the regulation pressure, the delivery valve is opened slowly. Similarly, the delivery valve is closed after stopping the power unit.

The pipes and sprinkler-lines are shifted as required after stopping. Dismantling of the installation takes place in the reverse order to the assembly.

Maintenance

General principles regarding the maintenance of the pipes and fittings and sprinkler heads are given below:

Pipes and fittings: The pipes and fittings require virtually no maintenance but attention must be given to the following procedures:

- Occasionally clean any dirt or sand out of the groove in the coupler in which the rubber sealing ring fits. Any accumulation of dirt or sand will affect the performance of the rubber sealing ring.
- Keep all nuts and bolts tight.
- Do not lay pipes on new damp concrete or on piles of fertilizer. Do not lay fertilizer sacks on the pipe.

Sprinkler heads: The sprinkler heads should be given the following attention:

- When moving the sprinkler lines, make sure that the sprinklers are not damaged or pushed into the soil.
- Do not apply oil, grease or any lubricant to the sprinklers. They are water lubricated and using oil, grease or any other lubricant may stop them from working.

Sprinklers usually have a sealed bearing and at the bottom of the bearing there are washers. Usually it is the washers that wear and not the more expensive metal parts.

Check the washers for wear once a season or every six months which is especially important where water is sandy. Replace the washers if worn.

After several season's operation the swing arm spring may need tightening. This is done by pulling out the spring end at the top and rebending it. This will increase the spring tension.

In general, check all equipment at the end of the season and make any repairs and adjustments and order the spare parts immediately so that the equipment is in perfect condition to start in the next season.

Storage

The following points are to be observed while storing the sprinkler equipment during the off season:

- Remove the sprinklers and store in a cool, dry place.
- Remove the rubber sealing rings from the couplers and fittings and store them in a cool, dark place.
- The pipes can be stored outdoors in which case they should be placed in racks with one end higher than the other. Do not store pipes along with fertilizer.
- Disconnect the suction and delivery pipe-work from the pump and pour in a small quantity of medium grade oil. Rotate the pump for a few minutes. Blank the suction and delivery branches. This will prevent the pump from rusting. Grease the shaft.
- Protect the electric motor from the ingress of dust, dampness and rodents.

Trouble Shooting

The following are the general guidelines to identify and remove the common troubles in the sprinkler systems:

- Pump does not prime or develop pressure:
- Check that the suction lift is within the limits. If not, get the pump closer to the water.
- Check the suction pipeline and all connections for air leaks. All connections and flanges should be air tight.
- Check that the strainer on the foot valve is not blocked.
- Check that the flap in the foot valve is free to open fully.
- Check the pump gland(s) for air leaks. If air leaks are suspected tighten the gland(s) gently. If necessary repack the gland(s) using a thick grease to seal the gland satisfactorily.
- Check that the gate valve on the delivery pipe is fully closed during priming and opens fully when the pump is running.

Sprinklers do not Turn

- Check pressure.
- Check that the nozzle is not blocked. Preferably unscrew the nozzle or use a

small soft piece of wood to clear the blockage. Do not use a piece of wire or metal as this may damage the nozzle.

- Check the condition of washers at the bottom of the bearing and replace them if worn or damaged.
- Check that the swing arm moves freely and that the spoon which moves into the water stream is not bent by comparing it with a sprinkler which is operating correctly.
- Adjust the swing arm spring tension, Usually it should not be necessary to pull up the spring by more than about 6 mm.

Leakage from Coupler or Fittings

The sealing rings in the couplers and fittings are usually designed to drain the water from the pipes when the pressure is turned off. This ensures that the pipes are automatically emptied and ready to be moved. With full pressure in the system the couplers and fittings will be effectively leak-free. If, however, there is a leakage, check the following:

- There is no accumulation of dirt or sand in the groove in the coupler in which the sealing ring fits. Clean out any dirt or sand and refit the sealing ring.
- The end of the pipe going inside the coupler is smooth, clean and not distorted.
- In the case of fittings such as bends, tees and reducers ensure that the fitting has been properly connected into the coupler.

Center Pivot Irrigation

The center pivot system consists of one single sprayer or sprinkler pipeline of relatively large diameter, composed of high tensile galvanized light steel or aluminum pipes supported above ground by towers move on wheels, long spans, steel trusses and/or cables. One end of the line is connected to a pivot mechanism at the center of the command area; the entire line rotates about the pivot. The application rate of the water emitters varies from lower values near the pivot to higher ones towards the outer end by the use of small and large nozzles along the line accordingly.

The center pivot (CP) is a low/medium pressure fully mechanized automated irrigation system of permanent assemble. It has become very popular in the Near East region in recent years for irrigation of most of field crops, cereals, legumes, forage and vegetables. It is also used for supplementary irrigation for rain fed grain. The cost of each system unit is relatively high and is therefore best suited to large irrigated farms. The area covered can be from 3.5 ha to 60 ha, according to the size of the CP, and the larger the area the lower is the cost of the system per unit area.

The center pivot.

System Layout and Component Parts

The typical center pivot system consists of a single long irrigating pipeline attached to a central tower and moves slowly over the field in a circular pattern and irrigates the plants with sprayers, or sprinklers placed on it at frequent spacing. The central tower with a pivot mechanism and main control panel (electric) is anchored to a small concrete base at a fixed water supply point (hydrant) at the center of the field. The entire irrigating pipeline is supported above ground by "A" frame towers move on wheels, long spans, steel trusses and/or cables; the end of the pipe is overhung with a sprinkler gun. The whole system rotates slowly, at a typical speed (last span) of 2–3 m/min., around the fixed pivot, self-propelled, applying water in the form of overhead spray irrigation and covers the area in a circular pattern. The drive system features small individual power units mounted on each wheeled tower. These units are electric drive, but can be hydraulic (water, oil) or mechanical drive. An automatic alignment system keeps always the irrigating pipeline straight.

The typical CP systems can be fixed permanent installations or movable/towable type with the central tower based on wheels or a skid, easily move from one field to another. The Linear Center Pivot is another common type towable system, which can irrigate rectangular or square shaped fields using a canal water resource parallel to the travel direction. Computer aided management systems (CAMS) models (MULTICENTER) and self-propelled are now available and the whole installation can be managed through a remote control for towing from one place to another. Corner systems are also available for irrigation of square, rectangular and odd shaped fields. Monospan systems are also available for small fields.

A center pivot overview.

Pipeline

The long irrigating pipeline (Lateral) with water emitters (sprinklers, bubblers, or sprayers) can be from 140 to 250 mm diameter, according to the system flow and the length; standard sizes of approximate 160 mm (6 inches) and 200 mm (8 inches) are very common. The length of the pipeline can be from 50 to 750 meters according to the design. It is made of high tensile material galvanized light steel, or aluminum, with extra strong couplers to stand the system operating pressures.

The pipeline is placed on wheeled "A frame" towers of typically 3 meters minimum height above ground and spaced 35–55 meters distance apart (length of spans). The common or "standard" length of spans is 40 m. Truss rod arches maintain the even distribution of weight and loads between the towers. On level ground the ground clearance varies from 2.75 to 4.5 meters for high profile machines. The spans are equipped with flexible joints at the ends allowing the pipeline to articulate and to allow side-to-side, up and down and rotational movement with no stress on the pipeline.

Water Emitters

The water emitters, computerized sized and spaced for high uniformity of application, are mounted on the pipeline at spacing of 1.5 to 3.0 m, and 6 m approx. according to the type and coverage of the sprayer emitters, and operate when the system is in motion. The emitters in the past were full- circle rotating sprinklers. Since the early eighties the Low Energy Precision Application (LEPA) mode is using sprayers, bubblers or angle mist sprayers, fitted on flexible "hose drops" hanging down from the lateral at a height above ground of about 20 to 45 cm for the bubblers and 1.0–1.8 m for the sprayers. The "hose drops" are connected to the pipeline by a "gooseneck" or furrow arm and operate at lower pressures of about 0.5 to 1.5 bars.

Goosenecks and drops are usually installed alternately on each side of the Lateral to even stresses on the line when used on high profile crops. There are several models of sprayers with excellent performance, long radius and uniform rain precipitation. Pressure/flow regulators are used in most cases. The discharge rate of the emitters along the pipeline is not the same along the line, but varies from lower values near the center to higher ones towards the outer end by the use of small and large nozzles along the line accordingly and sometimes variable spacing. Good overlapping is essential. Part circle sprayers are used near the towers to avoid over wetting the area along the wheels. The most common sprayers in use the last few years are the Senninger's wobblers and the Nelson's rotators.

A sprinkler gun is placed at the end of the overhang pipe and may increase the length of the line by of its sprinkling radius, yet under low pressures rotating sprinklers have poor performance. Operating pressures along the line are low, that makes the system sensitive to pressure changes caused from friction losses or differences in elevation and uneven ground. Installation of pressure gauges for frequent monitoring is important to ensure proper flow and good application efficiency and uniformity. A pressure gauge is always needed at the last drop of the pipeline. Since 0.4 Bars pressure regulators often used with LEPA need at least 0.65 bars at the inlet to function properly, then maintaining 0.65 bars at the end of the line will ensure that all regulators are operating properly. The inlet pressure needed for the CP normal operation is slightly less than 3 bars.

Low Energy Precision Application (LEPA irrigation) is defined as: a low pressure irrigation method for uniformity applying small frequent irrigations at or near the ground level to individual furrows (usually alternate furrows) with a mechanical move system accompanied by soil-tillage methods or tillage plus crop residue management to increase surface water storage capacity. It is used on-farm with small diameter bubblers located about 0.3 m above ground and with socks or sleeves discharging water directly into furrows.

Spray irrigation is defined as: the application of water by a small spray or mist to the soil surface, where travel through the air becomes instrumental in the distribution of the water. It was developed to reduce the droplet evaporation and drift inherent with impact sprinklers. The sprinkler manufacturers have developed many types of spray heads and today many combinations of pressure regulators and spray heads, nozzles and deflector plates are commercially available.

Today the CP irrigation systems are mainly available in three types: a) The LEPA sock, b) The LEPA bubbler, c) The sprayer.

The bubbler mode produces an umbrella shaped pattern approximately 0.4 to 0.5 m in diameter, which minimizes wind effects and only wet part o the soil surface. This so called true LEPA system as developed and introduced by Lyle and Bordovsky. The spray mode produces a horizontal spray approximately 2.5 to 3.3 m in diameter, which wets the entire soil surface. The so-called chemigate mode produces a 60° upward spray, which sprays the underside of the leaves). Lately new sprayers have been developed.

Hose drops with sprayers.

Commercial LEPA heads and nozzles, which can operate in the three different modes by changing the pad and hood positions, are now available. So, the same product can be used for pre-irrigation and germination, irrigation with the bubbler mode and chemigation.

Central Tower

This is a pyramidal structure of about 3.5–4.5 m height, built up with galvanized steel angular profiles and anchored on a concrete square platform. This structure has an access ladder. It is the head of the system and carries all equipment necessary for the control of the system, such as the system water fed up-going piece of pipe with the elbow on the top and inlets for fertilizer injection; the Collector ring, the Central Control Panel.

CP System Control

A modular control panel, protected in a cabinet, is installed on the pivot central tower and enables handling of the irrigation machine and the programming of irrigation, thus the control of flow and pipeline movement–operation time and speed/time per lap. A voltmeter and several pilot lights indicate control tension, support tower alarm and luck of pressure. Automatic starter, position stop device, automatic shut-off and hour counter are included too in a standard control panel. Manufacturers offer several models of various level of control including full telemetry control with the use of cellular telephone.

Sub-irrigation

Subirrigation also known as seepage irrigation, is a method of irrigation where water is delivered to the plant root zone. The excess may be collected for reuse.

Subirrigation is used in growing field crops such as tomatoes, peppers, and sugar cane in areas with high water tables such as Florida and in commercial greenhouse operations.

Three basic types of subirrigation system are in general use for potted plants in greenhouses: ebb-and-flow (bench-mounted enclosures holding pots are filled and then drained); trough (water is flowed through bench-mounted, slightly sloping enclosures containing pots); and flooded floor (special sloped concrete flooring is flooded and drained).

Greenhouse subirrigation has been growing in popularity since the 1990s. Advantages are water and nutrient conservation, and labor-saving. The outfitting cost is relatively high. Potential problems, such as the possibility of increased presence of disease in recycle water, have only begun to be investigated.

One of the disadvantages of sub-irrigated closed systems, such like Earth Boxes and sub-irrigated planters, is that soluble salts cannot be flushed into the lower soil profile and build up over time.

Hydraulic Agriculture

The adoption of hydraulic technology not only made agriculture physically less demanding on the workers, but also more efficient and with the growth in world population over the 20th century, the point is becoming increasingly important. While it may be too much to expect farmers to be expert engineers as well, it is useful to have at least a foundation in how hydraulic technology benefits agriculture.

Earlier hydraulic technology had fixed pressure and flow. Depending on the situation it was sometimes possible to change the settings, but this required human intervention. Modern tractors have a vastly greater hydraulic capacity than their predecessors with significantly more remote valves. At the same time, the traditional PTO shaft (complete with chains and drive shaft) has largely given way to hydraulic pipes and hoses which are driven directly by the hydraulic motor, making for simpler operation, meaning high reliability. As mentioned, the use of electronic controls allows for great precision, reducing wastage and therefore costs. In short, the combination of hydraulics and electronics has allowed agriculture to move on from a human worker making their best judgement about what is required to a machine using vast amounts of data from sensors to decide what is required and adjusting itself accordingly.

Key Priorities for Successful Agricultural Hydraulic Design

Energy Efficiency

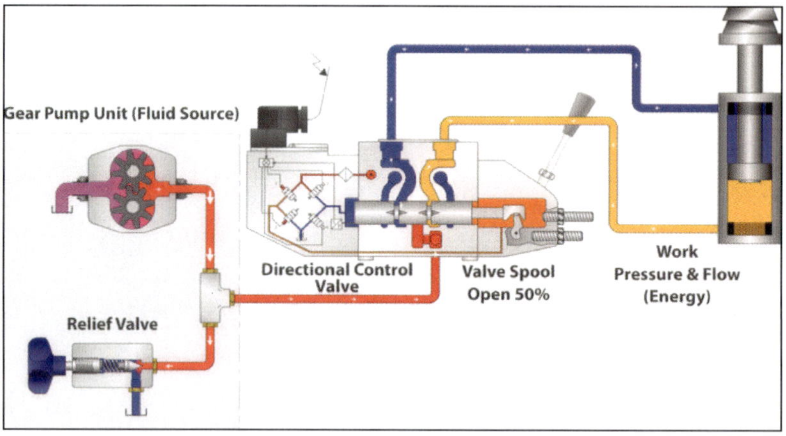

Use of a gear pump typically means that system pressure is at maximum (relief valve setting) when using only a portion of total available flow to control cylinder speed.

Assuming that hydraulic motor torque requirements and displacements have been calculated and that cylinders have been sized to produce the required lifting forces (within a normal range of hydraulic pressures), the choice of pump size and style will be the next major decision. For a small scale farm, a utility type of machine may incorporate a fixed displacement gear pump with an open center valve bank for the sake of simplicity and low cost. Even though the pump may be pressure-unloaded with flow directed back to tank when no hydraulic function is in use, higher-than-needed fuel consumption still occurs due to the full flow rate.

When only part of the total flow available from the pump is needed for a cylinder—deliberately extended at slow speed by judicious use of the valve lever—the excess flow is forced over a spring-loaded pressure relief valve. The partially shifted valve spool closes the pump unload path to tank, and places a partially open valve in the flow path as a major restriction and pressure load. The pump is forced to work at full flow and at maximum pressure, even if the cylinder is only lifting a light load. In other words, the flow requirement for cylinder speed may only be a fraction of the total flow available, yet the pump can only pump at the full rate. In addition, the cylinder may only need to lift a light load (requiring minimal pressure), yet the system overall will be at maximum pressure due to the restriction of the partially opened directional valve. Fuel consumption is highest when pumping at full flow and full pressure, but this may be acceptable for a simple tractor with manual steering, and where hydraulic cylinder motion is occasional while working with the machine.

For large, production-scale machinery with continuously running hydraulic functions such as planting or harvesting circuits, and with the need for forceful steering actions at any time, the variable displacement piston pump remains the standard choice. A pressure-compensated and load-sensing piston pump avoids the inefficiencies of a simple gear pump. It provides a flow rate that automatically matches cylinder and hydraulic motor flow demand, across multiple parallel circuit paths, and also sets the maximum system pressure only slightly higher than what is currently needed for the highest operating load.

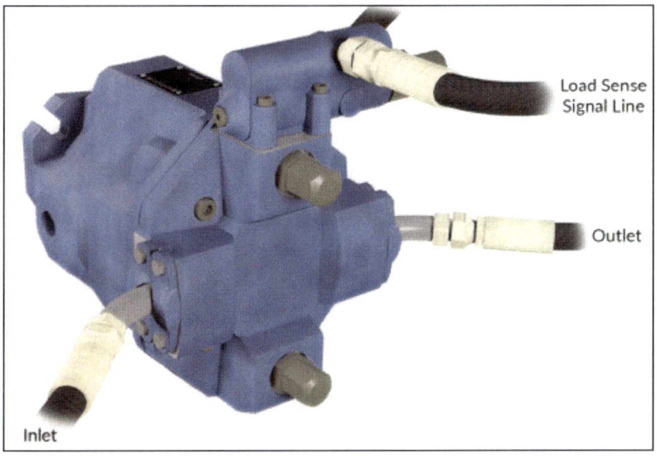

Pressure feedback is delivered to the load sense compensator on the pump from the highest pressure circuit within the system. This feedback adjusts the pump on the fly. With the maximum pump displacement sized just slightly larger than the anticipated maximum flow, fuel efficiency is delivered by a pump that can continuously adjust pressure and flow as needed.

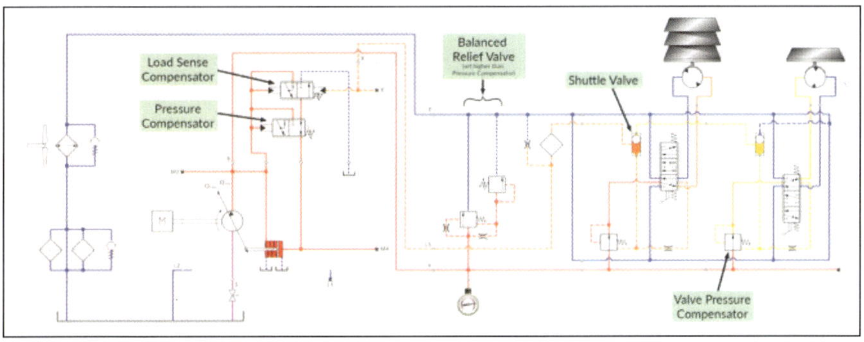

Schematic of typical load sensing (flow compensating) hydraulic system. Brake-shoe, brick stacks only for ease in understanding load pressures.

The load sensing feature also offers greater flow accuracy within the system overall. Minor changes in engine speed do not show up as flow variances because the load sense compensator can adjust the pump automatically. If the hydraulic system is operating motors on timed production functions on a harvester, then this flow accuracy has a direct impact on production and quality.

The pressure compensator is a control mechanism on the pump that reacts when a preset maximum pressure level is reached. Instead of using a system relief valve as the primary means of pressure limiting, the compensator reduces the pump displacement and keeps it near zero until the excessive pressure condition (e.g. an overload at cylinder or motor) dies down. When pressure drops down, the compensator allows the pump to increase displacement once again. The pressure compensator is similar to a standard relief valve in that it keeps the system pressure from rising above a safe level. But the pressure compensator does this by reducing the pump output, rather than inefficiently diverting extra flow back to tank. The pressure compensator provides a similar safety feature as a relief valve, but with the added bonus of energy savings.

The flow and directional control valve banks on most modern tractors and harvesters feature electronic control. With variable current solenoids, the valves are capable of proportional flow to set cylinders and motors in motion at the desired speed.

The valves can hold a cylinder in a locked position (e.g. crop cutting height or plow depth), or allow a cylinder rod to move freely in and out so that an implement can float up and down with ground contours. For smaller, directly attached implements, this raise-and-lower cylinder is found on the tractor's three-point hitch. For larger towed implements with their own cylinders and hydraulic motors, pairs of quick couplers from tractor valves commonly referred to as "remotes" are found at the rear of the tractor to receive hydraulic hose connections.

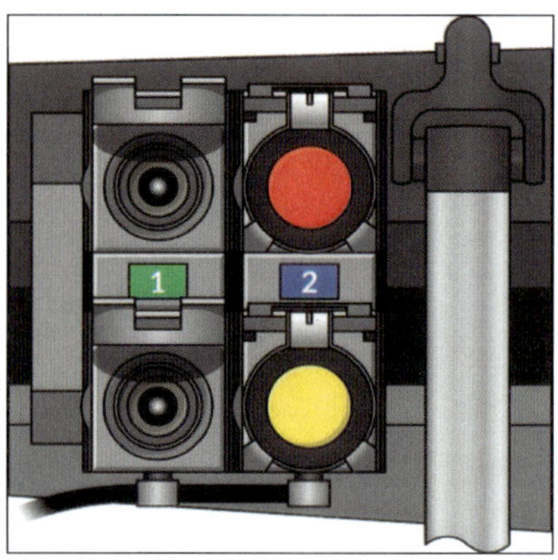

While these look like basic connections to a source of flow, these remote valve connections bring a level of sophistication to simple hydraulic functions on towed implements. The valve bank lurking behind modern remote couplers can be capable of high levels of automation.

A pressure compensator is typically found within each section of the valve bank. Most often this extra valve component is found on the inlet side of the main spool. The compensator senses the load pressure of the cylinder or motor in motion, and adjusts the pressure at the inlet of the main valve spool to ensure a steady pressure differential across the spool, thus maintaining a steady flow rate even when system pressure or cylinder load pressure changes. This means that the operator does not have to quickly move a valve lever in reaction to a hydraulic function that is changing speed during motion. The pressure compensator eliminates the problem.

The valve bank contains a network of shuttle valves with tiny steel balls or poppets inside. These shuttle valves (aka ball resolvers) allow for the load pressure from only the highest loaded valve section to be transmitted to the load sense compensator on the pump, via a special signal hose. The use of the load sense compensator on the pump and the valve bank with shuttle valves as described completes a closed loop control scheme with the use of a fluidic pressure line to deliver feedback from the load to the pump.

Load-sensing systems are considered to be among the most fuel efficient for continuously running hydraulic systems. They can also be challenging systems to correctly adjust for tractors pulling implements of infinite design variation. There are lots of components involved, making system troubleshooting a complex process.

Recent innovations such as Parker Hannifin's Split Pump Intelligent Flow Architecture use one piston pump for each major function on the machine, allowing for even greater energy savings. These types of systems, already proven for loaders and some excavators, will slowly make their way onto agricultural machines. The pump flow rate and the proportional valve spools are electronically controlled. This offers options around combining pumps for use with one implement cylinder during fast motion requirements, or for separate functions when operating several different cylinder motions at the same time. The internal parts of the valve bank are simpler and the use of a load-sense signal hose between pump compensator and the valve bank is eliminated in favor of faster electronic pump control.

Energy saving designs for hydraulic systems on large agricultural machines also translate to lower oil temperatures, which lower or even remove the need for large coolers and fans. A size reduction for fluid reservoirs is also possible for these systems.

Ease of Operation

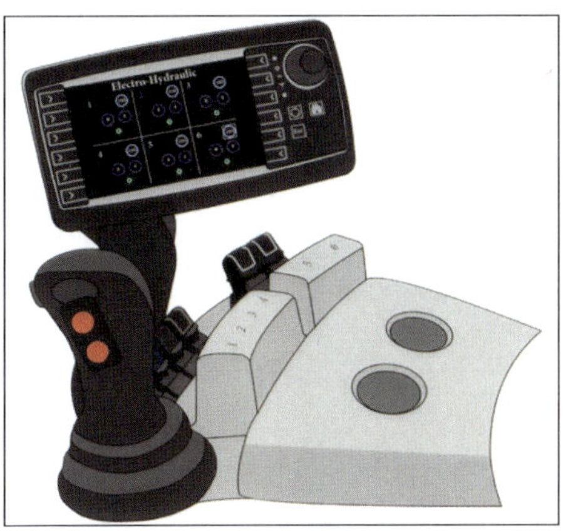

Touchscreen programming of timed hydraulic functions and finger-touch valve switches provide convenient automation options.

Hydraulic functions on an agricultural machine are considered easy to operate when the operator does not have to make continuous corrections to cylinder positions (e.g. crop cutting height, steering) or adjust flow rates to hydraulic motors. Quality and production for plowing, planting or harvesting are the top priorities on large-scale machines. The operator's brainpower is best used to monitor the big farming picture. With the high travel speeds and the wide crop rows possible on high horse-power machines, the operator still has lots to do. Operator-programmed controllers are extremely common. These controllers can be programmed to remember how much time it takes for hydraulic cylinders to unfold an implement. This "timed detent" feature can then be used later with the push of a button to accomplish the same operation.

The minor points of machine control and position correction should be automatic with an allowance for operator override. Automation and operational ease is the result of typical electronic controller and touch screen use with electrohydraulic valves. The possibilities for a machine that in the past was hard physical work to operate, continue to expand with steering guidance systems now moving well beyond the servo-operated steering wheel. Unlike the world of construction machinery, where steering is often controlled by joystick on many models, the familiar steering wheel connected to an orbitrol valve (aka hand metering unit) is slow to disappear from agricultural machines. There are however, entirely automated steering controls and valves connected in parallel to the steering wheel on some machines.

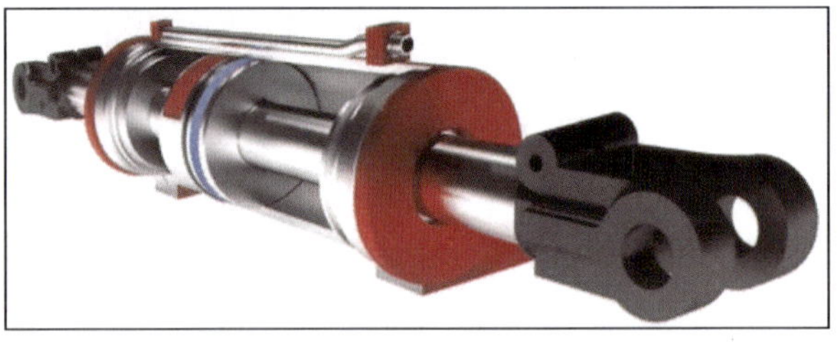

Modern sensors, such as this Hall effect, cylinder position sensor shown, offer new automation possibilities.

Designers should now give standard consideration to the installation of linear position sensors inside the cylinders on the steering axle or in the cylinders used for articulated steering. Non-steering cylinders are also worth considering for the installation of linear position sensors, to enable more position memorized automation. Some newer sensors that are suitable for double rod steering cylinders can mount on the outside of the cylinder tube with only the magnet inside on the piston. The electronic control equipment on the machine may not yet be configured for fully automated, robotic steering, but many systems already offer a level of GPS-based guidance for the long straight runs across the field, allowing the operator to avoid fatigue.

Ease of Maintenance and Repair

If you were in charge of engineering this hypothetical crop harvester, where would you place the filter? Behind the wheel gives a cleaner look, but out in the open makes the filter accessible for routine maintenance. Small design decisions like this one can have large consequences.

Ease of maintenance and repair is a valued feature that many farm operators look for. While easy-to-access fill ports and easy-view sight glasses are welcomed, many speak with notable praise for the machine designer who makes adjustments and wrench work easy. With 45° elbow fittings, or wide radius tube elbows at the cylinder ports, the connections are easy to work on when damaged hoses have to be replaced. Rows of closely spaced hoses and fittings clamped along the machine frame can work out sufficiently if the fittings are offset from each other to allow room for wrenches. Mounting filters for easy removal and change-out is not always the most visually pleasing layout, but mechanics appreciate it. It also helps to make sure that maintenance tasks are not skipped.

Even something as basic as filter placement can have a big impact on maintenance; if filters are accessible and easy to remove/change out, this simple maintenance task is less likely to be overlooked.

Agricultural machinery continues to be a key product for the application of hydraulic systems. As innovative components and system designs that make for more efficient operation are being released from time to time, the use of automation is increasing very rapidly. This is especially true for large-scale machines designed to help the farm operator with production yield, time usage and human input energy.

It is exciting to see innovations from other industries, such as mining and construction, crossing over into the agricultural world and making a difference in the average farmer's day. As innovations in hydraulic system design continue to filter into agriculture, farmers will enjoy the rewards of even greater efficiency, ease-of-operation, and hopefully, reductions in maintenance complexity.

Usefulness of Hydraulics in Agriculture

Hydraulics is literally feeding the world. However, like health and freedom, the importance of hydraulics technology becomes apparent when absent. In the past, agricultural workers often went through grueling work, but their work today is physically less demanding, thanks to the adoption of and advances in hydraulic technology. The contribution that this technology has made towards improving agricultural practices is immeasurable.

Hydraulic Technology

When hydraulic technology was adopted to replace horses, there was hope that agricultural production would increase. The application of hydraulics in agriculture started in earlier tractors, which used hydraulics internally for steering and braking of the equipment and externally for heavy lifting and tedious digging to prepare land for planting. Despite the need for human intervention, the advent of hydraulic technology has helped to reduce reliance on humans to plough land, perform weeding, and harvest crops.

Agricultural workers and experts would agree that hydraulics in agriculture has advanced significantly from performing simple tasks such as lifting to more complex power implements. Agricultural technology has advanced to a level where machine operators do not have touch a button (switch) or move a lever to facilitate the flow of liquid in the direction required, as tiny computer processors act on pumps and valves by giving out commands at faster speeds than the human mind can think and process.

Modern-day Hydraulic Systems

The technologies allow machine operators to accomplish tasks that enable farmers to produce sufficient food to feed the population, at competitive prices. Without advanced technology and applications of hydraulics in agriculture, prices of food would remain dramatically high, leaving many people unable to afford. In has increasingly become important for tractor owners and farmers to understand the technology behind hydraulic systems so that safety is prioritised when operating them. Being informed of current technology used in hydraulic systems will enable them to take full advantage of added features to maximise their output or benefit from the machinery and equipment in numerous ways.

References

- "Self-driving Ibex robot sprayer helps farmers safely tackle hills - Farmers Weekly". Farmers Weekly. Retrieved 2016-03-22
- Mechanization-of-agriculture-meaning-benefits-and-progress, farming, india: economicsdiscussion.net, Retrieved 19 April, 2019
- Pedersen, S.; Fountas, S.; Blackmore, S. (2008). "Agricultural Robots – Applications and Economic Perspectives". In Takahashi, Yoshihiko (ed.). Service Robot Applications. Rijeka: intech. Pp. 369–382. Doi:10.5772/6048. ISBN 978-953-7619-00-8
- "Machinery Manufacturers - Secondary Tillage". Worldwide Agricultural Machinery and Equipment Directory. 2006-11-14. Retrieved 2007-07-12
- Wendel, Charles H. (2004), Encyclopedia of American Farm Implements and Antiques (2nd ed.), Iola, WI, USA: Krause Publications, ISBN 978-0873495684
- Belton, Catherine; Farchy, Jack; Blas, Javier (5 August 2010). "Russia grain export ban sparks price fears". Financial Times. Retrieved 6 June 2014

- Ahmadi, H.; Golian, A. (2010-11-01). "The integration of broiler chicken threonine responses data into neural network models". Poultry Science. 89 (11): 2535–2541. Doi:10.3382/ps.2010-00884. ISSN 0032-5791. PMID 20952719
- "Satellite monitoring data use // Current use and potential of satellite imagery for crop production management. B. De Solan, A.D. Lesergent, D. Gouache, F. Baret". Europa.eu. Retrieved 10 January 2018
- Application-of-Drone-in-Agriculture: researchgate.net, Retrieved 5 February, 2019

Dairy Farming Technology

5

- **Artificial Insemination**
- **Mixer-wagon**
- **Grain Crimping**
- **Milking Pipeline**
- **Bulk Tank**
- **Automatic Milking**

The class of agriculture which deals with the long-term production of milk is referred to as dairy farming. Some of the modern techniques and technologies used for dairy farming are artificial insemination, grain crimping, automatic milking, etc. The diverse applications of these dairy farming technologies have been thoroughly discussed in this chapter.

Technology has greatly enhanced dairy farming processes making them easier and simpler. This, in turn, has lowered dairy farming costs as compared to other years. Currently, the number of dairy products in the market has greatly multiplied as a result of advancement in dairy farming. Therefore, advancement in technology has affected dairy farming positively.

Latest technology in dairy farming performs a number of functions like recordings daily milk yields, monitoring of milk component such as protein and fat, pedometers, recording temperature automatically, indicating milk conductors, detecting of estrus in an automatic manner and finally taking daily measurements of the animals' body weights.

The respective kind of technologies has greatly caused impacts in quality, healthy and reproduction control. The major benefits associated with these are reduced costs, improved quality of products, increased efficiency, reduction of the adverse environmental effects and improved performances in the health of dairy animals and their general well-being.

Examples of Latest Technologies in Dairy Farming

Technology has enhanced the growth of dairy cattle in a correct and healthy manner. In addition, it has helped in boosting of milk production and monitoring of these herds.

For instance, the use of drone technology has been effectively used to monitor dairy animals. This is because the drone works to ensure that everything is in its right location. It makes this possible by flying over the properties.

Another important form of latest technology in dairy farming is the issue of computer farming system. This system works by using facial recognition software to monitor cows. This is vital in helping to give out physical information including the invisible one.

Also, other dairy farmers nowadays have adopted the technology of robotic cow milking equipment which does the work repetitively and precisely. The latest technology is applied here to make the work effective every time for the purpose of attaining and maintaining the required standards.

There is also another important latest device used in dairy farming called the Delavan swinging cow brush. This play a major role in ensuring that the cattle are comfortable during their rotations. The brush helps to stimulate blood circulation which keeps the cows healthy and comfortable always leading to a high quantity of milk production.

Moreover, dairy farmers have adopted a way of keeping track of the cow's health, milk production levels and the number of cows which have already been milked. This is made possible through the use of sensor technology which provides information to the farmer through a transporter.

Furthermore, nowadays farmers access apps which have built up and made their operations even easier as compared to other periods. This app allows them to get access to skills that are so essential in addition to finding the potential market for the products mostly online markets.

Currently, farmers are looking for alternatives to power sources as opposed to electricity as one way of making farming operations easier than before. This, in turn, will promise continuous production of dairy products. One way of making this process successful is through the use of solar panels to generate power to their homes.

Finally, another form of latest technology in dairy farming is the use of dairy shed technologies. This one has been used effectively in increasing the quantity of production. For instance, the use of automated cup removers and spraying and automated drafting. Also the introduction of automated mastitis detection system which plays a vital role in making detection of diseases.

Artificial Insemination

As the name suggests, Artificial insemination (AI) is a technique in which sperm is collected from the male/bulls, processed, stored and manually introduced into the female reproductive tract at appropriate time for the purpose of conception. AI has become one of the most imperative techniques for the genetic improvement of farm animals since preferably the semen from genetically superior sires/males are used to inseminate the female animals artificially. It has been most extensively used for breeding dairy cattle and buffaloes. AI is the tool that possesses the potential of economic and rapid diffusion of elite males to a large number of females in a short duration of time over a large geographical area.

Advantages of Artificial Insemination over Natural Service

Artificial insemination (AI) is one of the most efficient tools accessible to dairy farmers to improve productivity and profitability of dairy enterprise. In artificial insemination the bulls of superior quality can be efficiently exploited with the least concern for their location in faraway places. There are a lot of advantages of AI over natural services with bulls. They are as follows:

- *Boosts efficiency of bull usage*: During natural mating, a bull will donate much more semen than is theoretically needed to make a pregnancy. On the other hand, collected semen can be diluted and extended to make hundreds of semen doses from a single ejaculate which can be easily carried one place to another, promoting multiple inseminations in females in different geographical locations and semen can be stored for long periods of time

- *Cost Effectiveness:* No necessity of maintenance of breeding bulls. Hence, the expenditure on maintenance of breeding bull is saved.

- *Checks disease transmission:* Natural mating allows the transmission of venereal diseases between males and females. On the other hand, for AI, semen is regularly tested for its quality, possible infections hence allows checking of the spread of certain venereal diseases. Eg: contagious abortion, vibriosis.

- *Promotes Breeding Efficiency:* By routine examination of semen after collection and frequent checking on fertility make early detection of inferior bulls and better breeding efficiency is warranted.

- The progeny testing can be employed at an early age.

- The semen of an elite bull can be used even after the death of that sire.

- It makes possible the mating of animals with great variations in body size with no injury to either of the animal.

- It is useful to inseminate the cows denying to stand or accept the bulls at the time of oestrum.

- Useful in maintaining the perfect breeding and calving records.

- Artificial Insemination enhances the rate of conception.
- Artificial Insemination allows the use of old, heavy and injured sires.
- Artificial Insemination when linked to oestrous synchronization programme, can promote a more consistent, uniform calf crop production.

Shortcomings of Artificial Insemination

Dairy farmers have numerous arguments against Artificial Insemination. In general, an AI program would need more intensive management of the herd like a sound nutrition program (cows in good body condition), good record keeping in the farm, an efficient herd health program, precise heat detection and a well-trained AI technician. Poor management in one or more of these fields might result in poorer success rates. Few major drawbacks of AI are mentioned below:

- Requires well-trained personnel and special tools.
- Takes more time than natural services.
- The AI operator needs to have the knowledge of the structure and function of reproduction in the dairy animals.
- Improper cleanliness of instruments and unhygienic conditions may lead to declined conception.
- If the bull is not appropriately examined, the spreading of venereal diseases will be increased.

Mixer-wagon

A mixer-wagon, or diet feeder, is a specialist agricultural machine used for accurately weighing, mixing and distributing total mixed ration (TMR) for ruminant farm animals, in particular cattle and most commonly, dairy cattle.

A paddle-type mixer-wagon coupled to a tractor.

A weighing computer showing an empty mixer-wagon, 0 kg tare.

Trailed mixer-wagons vary in size from 5 m^3 to more than 45 m^3. Some self-propelled mixer-wagons may be bigger than this. Displacement varies according to ration dry matter. More water means more weight. With dry (45% DM; dry matter) rations, a 14m^3 mixer-wagon such as the one pictured, may contain 3 tonnes fully loaded, or enough for about 60 Holstein cows.

The paddle turning inside a mixer-wagon.

Stationery knives are fitted on the side of the tub.

A mixer-wagon commonly consists of:

- A trailed chassis, for coupling to a tractor power unit, and fitted with one or more usually braked axles, and fitted with a road-legal lighting system.
- A mixing body, attached to the chassis by four weighing sensors, one at each corner. There are three main types of mixing body:
    - Paddle, whereby a central axial shaft turns a series of paddles which rotate the contents and provide front to back mixing in the mixer-wagon.
    - Vertical auger, with one, two or three augers used to move the contents from top to bottom.
    - Horizontal auger, containing between one and five augers, used to circulate the contents from front to back and bottom to top.
- A digital weighing computer working through the above-mentioned weighing sensors. Such a computer can typically memorise 9 or more different rations and 99 or more ingredients.

The system can weigh in all the ingredients of a ration in any chosen order and weigh out according to chosen quantities corresponding to the needs of different groups of animals. As each ingredient is loaded, a visual and audible signal alerts the operator when the required amount is reached. The flashing light and "beep" system repeats faster and faster until the ingredient is complete, when the signal becomes continuous for 2 seconds. The computer then shifts to the next ingredient and displays the product name and weight to be loaded. Loading is generally done using the mixer-wagon cutting and loading device, a high capacity tractor loader, or a telescopic handler.

- The mixing paddle, rotors or augers are connected to the tractor pto through a reducer system, provided by either a planetary gearbox, and/or a step-down pulley and chain system.
- A set of stationary knives, against which long fibre may be chopped by the forcing action of the mixing rotor.
- A hydraulic door to seal the ration in during mixing, thus permitting the use of liquid feeds such as molasses.
- An unloading system, consisting of a simple hydraulically adjustable chute, up to a hydraulic powered conveyor belt.
- Self-propelled mixer-wagons are mounted on a lorry chassis or may be specialist self-loading machines.

Grain Crimping

Grain crimping or moist grain crimping is an agricultural technology, an organic way to preserve feed grain into livestock fodder by fermentation.

Crimped grain brings health benefits to the animals and economic benefits such as cost savings and increased meat or milk production to the farmer.

Crimping was developed in Finland in the end of the 1960s by two farmer brothers, Aimo and Gunnar Korte, based on findings of British researchers as early as 1918. The brothers made the first crimper machine for home farm use, and as the word started to spread, they set up a company to manufacture and sell the machines.

The British researchers had established, that grain attains its peak nutritional value when the moisture content of the grain is between 35% and 45%. However, it took about 50 years before this knowledge was successfully turned into a method to process and preserve the grain when still moist.

Traditionally, grain is not harvested until it is dry enough to be ground by a hammer mill. Moist grain often cannot be ground or stored without machine drying and using preservatives, which always increases costs.

In crimping, the grain is combined moist and run through the crimper machine, which will break and flatten the grains. Additives, such as certain preservatives or molasses and water (if necessary) can be added in order to ensure the protection of nutrients.

Crimped grain is stored in storage silos as a silage.

Crimped grain is dustless, thus convenient to handle, does not require any further processing, and is often preferred by the animals to drier and dustier feeds.

Practical experiments by farming and livestock research institutions in Finland, Sweden, UK and elsewhere have confirmed, that crimped feed has higher nutritional values, it increases the animals' growth and milk production, improves milk quality and the animals' health, and in addition, helps cut costs.

An important point is that crimping home-grown grain and processing the feed on the spot at the farm, the feed ingredients can be controlled and are fully traceable, thus helping in prevention of diseases, such as Bovine spongiform encephalopathy.

Milking Pipeline

A milking pipeline or milk pipeline is a component of a dairy farm animal-milking operation which is used to transfer milk from the animals to a cooling and storage bulk tank.

In small dairy farms with less than 100 cows, goats or sheep, the pipeline is installed above the animals' stalls and they are then are milked in sequence by moving down the row of stalls. The milking machine is a lightweight transportable hose assembly which is plugged into sealed access ports along the pipeline.

In the United States, for farmers who participate in the voluntary Dairy Herd Improvement Association, approximately once a month the milk volume from each animal is measured using additional portable metering devices inserted between the milker and the pipeline.

In large dairy farms with more than 100 animals, the pipeline is installed within a milking parlor that the animals walk through in order to be milked at fixed stations. Because the machine is stationary, it can include additional fixed equipment such as computerized milk-metering systems to measure volume, which would be cumbersome to use with portable milkers.

In both cases the pipeline is constructed out of stainless steel, which does not easily corrode and is resistant to most chemicals, though larger operations may use larger-diameter pipes in order to handle greater milk volumes.

Transfer from Pipeline to Bulk Tank

There is usually a transition point to move the milk from the pipeline under vacuum to the bulk tank, which is at normal atmospheric pressure. This is done by having the milk flow into a receiver bowl or globe, which is a large hollow glass container with electronic liquid-detecting probes in the center. As the milk rises to a certain height in the bowl, a transfer pump is used to push it through a one-way check valve and into a pipe that transfers it to the bulk tank. When the level has dropped far enough in the bowl, the transfer pump turns off. Without the check valve, the milk in the bulk tank could be sucked back into the receiver bowl when the pump is not running.

In the event of electronics or pump failure, there is also usually a secondary bowl attached to the top of receiver bowl, which contains a float and a diaphragm valve. If the main receiver bowl overflows due to pump failure, the rising milk lifts the float in the secondary bowl, which will cut off vacuum to the entire milk pipeline and will prevent the milk or wash water from being sucked into the vacuum pump.

Some milk handling systems eliminate the receiver bowl and transfer pump by having rubber seals on the bulk tank covers, to permit the entire tank to be under vacuum until milking is finished. Milk can then just flow directly by gravity from the pipeline into the bulk tank.

Pipeline Cleaning

The pipeline and all milk handling systems are cleaned after every milking session using a washing system that first rinses out the remaining milk and then flushes cleaning solution through the piping to kill bacteria and remove *milkstone*, a layer of scale mainly formed by cations like calcium and magnesium. The entire washing mechanism is operated very much like a household dishwasher with an automatic fill system, soap dispenser, and automatic drain opener.

The pipeline is usually set up so that the vacuum in the system that lifts milk up can also be used to drive the cleaning process. Rather than having a single line run to the bulk tank, typically a pair of lines transport milk by gravity flow to the receiver bowl and transfer pump. The high ends of these two lines are joined together to form a complete loop back to the receiver bowl.

Cleaning is accomplished by inserting a choke plug into one of the lines leading to the transfer pump, and sucking large volumes of water from a wash-water supply tank into the choked line. This choke plug is mounted on a rod, and is inserted into the line before cleaning, and pulled out for regular milking. Due to the choke, the water, which is sufficient to completely fill the pipe, is sucked up one side of the pipeline, over the high point joining the two pipeline sections, and then flows back to the receiver bowl and transfer pump through the unchoked line. The transfer pump is then used to move the cleaning solution from the receiver bowl back to the wash-water supply tank to restart the process.

Typically, the inlet ports on the receiver globe are designed so that large slugs of wash water moving at high speed will enter on a tangent to the sides of the globe and rapidly spin around inside to assist in vigorous cleaning of the globe's interior. It is normal for wash water to overflow out the top of the globe and for some wash water to be sucked into the overflow chamber to also flush it out. During cleaning the bottom of the overflow chamber is connected to a drain channel on the receiver globe to permit water to flow out.

For the small-farm pipeline, portable milkers are inserted into this cleaning loop usually by sucking the cleaning solution out of the wash supply tank through the milker claw and outputting from the milker hoses into the choked end of the line. When the water returns to the receiver bowl, the transfer pump returns the water back to the milker's water pickup tank.

Bulk Tank

In dairy farming a bulk milk cooling tank is a large storage tank for cooling and holding milk at a cold temperature until it can be picked up by a milk hauler. The bulk milk cooling tank is an important piece of dairy farm equipment. It is usually made of stainless steel and used every day to store the raw milk on the farm in good condition. It must be cleaned after each milk collection. The milk cooling tank can be the property of the farmer or be rented from a dairy plant.

Bulk Tank Types

Different types of milk cooling tanks.

Raw milk producers have a choice of either open (from 150 to 3000 litres) or closed (from 1000 to 10000 litres) tanks. The cost can vary considerably, depending on manufacturing norms and whether a new or second hand tank is purchased.

Milk silos (10,000 litres and plus) are suitable for the very large producer. These are designed to be installed outside and adjacent to the dairy, all controls and the milk outlet pipe being situated in the dairy.

Tank Construction

A milk cooling tank, also known as a bulk tank or milk cooler, consists of an inner and an outer tank, both made of high quality stainless steel.

The space between the outer tank and the inner tank is isolated with polyurethane foam. In case of a power failure with an outside temperature of 30 °C, the content of the tank will warm up only 1 °C in 24 hours.

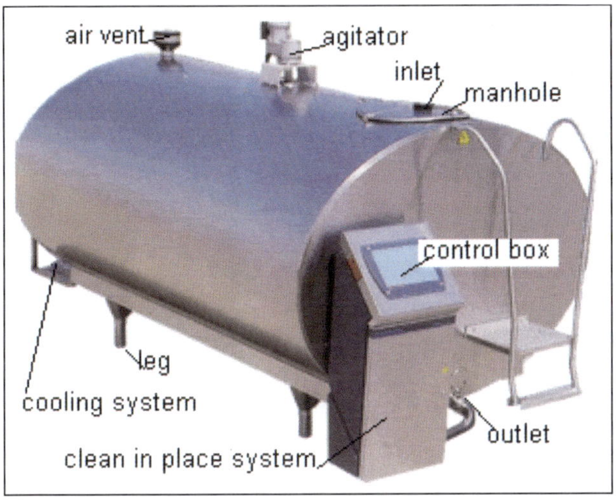

Bulk milk cooling tank.

To facilitate an adequate and rapid cooling of the entire content of a tank, every tank is equipped with at least one agitator. Stirring the milk ensures that all milk inside the tank is of the same temperature and that the milk stays homogeneous.

On top of every closed milk cooling tank is a manhole of about 40 centimetres diameter. This enables thorough cleaning and inspection of the inner tank if necessary. The manhole is covered by a lid and sealed watertight with a rubber ring. Also on top are 2 or 3 small inlets. One is covered with an air-vent, the other(s) can be used to pump milk into the tank.

A milk cooling tank usually stands on 4, 6, or 8 adjustable legs. The built-in tilt of the inner tank ensures that even the last drop of milk will eventually flow to the outlet.

At the bottom, every milk cooling tank has a threaded outlet, usually including a valve.

All tanks have a thermometer, allowing for immediate inspection of the inner temperature.

Most tanks include an automatic cleaning system. Using hot and cold water, an acid and/or alkaline cleaning fluid, a pump and a spray lance will clean the inner tank, ensuring an hygienic inner environment each time the tank is emptied.

Almost every tank has a control box. It manages the cooling process by use of a thermostat. The user can turn the system on and off, allow for extra and immediate stirring, start the cleaning routine, and reset the entire system in case of a failure.

New and bigger milk cooling tanks are now being equipped with monitoring and alarm systems. These systems guard temperature of the milk inside the tank, check the functioning of the agitator, the cooling unit and temperature of the cleaning water. In case of malfunctioning of any of these functions, the alarm will activate. The monitoring system will also keep a record of the temperature and of all malfunctions for a given period.

Bulk Tank Manufacturing Norms

Norms define among other criteria: insulation, milk agitation, cooling power required, variations in milk quantity measurement, calibration, Some are more demanding than others.

- ISO standard 5708 (Refrigerated bulk milk tanks);
- European standard EN 13732 (*Food processing machinery* – Bulk milk coolers on farms – Requirements for construction, performance, suitability for use, safety and hygiene).

Cooling Systems

There are two primary methods of cooling milk entering the bulk tank, each with its own advantages and disadvantages. The tank capacity and type will depend on herd size, calving pattern, frequency of milk collection, required milk quality, energy and water availability and future plans for development.

Direct Expansion

A bulk tank with direct expansion cooling has pipes or pillow plates carrying refrigerant which are welded directly to the exterior of the milk chamber. A layer of insulation covers the exterior of the milk tank and the cooling lines, with an exterior metal shell over the insulation.

Direct expansion cooling cannot run when the tank is empty or the inside walls of the tank would freeze. Instead, the tank is rapidly cooled as warm milk first enters the tank, and then the tank is cooled slowly just to maintain a low storage temperature. The rapid cooling during milking requires very large refrigeration compressors and condenser radiators to quickly expel heat from the milk, and is better suited for very large farming operations where three-phase electric power is available to operate the high-power cooling system.

Ice Bank

A bulk tank using an Ice Builder or Ice Bank immerses the bottom of the inner milk chamber in an open pool of water with copper tubes containing refrigerant suspended in the water. Between milkings, a small low-power cooling system slowly builds up a coating of ice around the copper tubes, and prevents icing of the pool over by continuously circulating the water in the pool. After the ice has achieved a thickness of 2-3 inches, the cooling system stops running.

During milking, the milk entering the tank is primarily cooled by circulating the water in the pool around the walls of the inner milk chamber, and the melting of the ice. After the ice has melted sufficiently the cooling system restarts to assist the ice bank and restart the ice building.

Ice bank bulk tanks are better suited for small family farm operations where only single-phase electric power is available, and high-power cooling systems would be either too expensive or difficult to install.

Milk Pre-cooling

For energy savings and quality reasons it is advisable to pre-cool the milk before it enters the tank using a plate or a tube cooler (shell and tube heat exchanger) supplied with chilled water from the well water, the ice builder or the condensing unit. The quicker milk is cooled after leaving the cow the better. This system achieves most of the cooling before the milk enters the tank, so that chilled milk, rather than warm milk, is being added to the already cooled milk in the tank.

Cooling Temperature

Generic temperature for milk storage is 3 to 4 °C. For raw milk cheese manufacturing, it would be advisable to keep the milk at 12 °C, as milk characteristics will be kept in a better state.

The milk cooling tank is usually not completely filled at once. A 2 milking tank is designed to cool 50% of its capacity at once. A 4 milking tank is designed to cool 25% of its capacity at once, and a 6 milking tank is designed to cool 16.7% of its capacity at once.

The cooling performance depends on the number of milking it takes to completely fill the tank, the ambient temperature and the cooling time.

Bulk Tank Cleaning Systems

There are two primary methods of cleaning bulk tanks, via manual scrubbing or automatic washing. Both methods generally use four steps to clean the tank:

- Prerinsing with water to wet the surface and rinse off remaining milk residue;
- Washing with hot soapy water;
- Rinsing with water to remove the soap;
- Final *sanitizing* rinse with an approved bulk tank sanitizer solution.

Manual Scrubbing

Manual scrubbing requires the bulk tank to have large hinged covers that can be lifted open to permit easy access to the interior surfaces of the tank. It tends to be much more thorough than automatic methods since it permits the tank to be carefully inspected during the washing process. If the tank is not found to be cleaned well enough, a troublesome area can be given additional cleansing attention.

Manual Scrubbing Limitations

This job is difficult to perform for very large tanks, and becomes more difficult as the overall cross-section or diameter of the tank increases, requiring either a longer brush or a raised work platform around the tank to lift the cleaning worker to reach over the side of a tall tank.

Automatic Washing

Automatic bulk tank washing and are normally activated by the milk collection truck driver after each milk collection. The cleaning system operates similar to a consumer dishwasher and consists of one or more free-spinning high-pressure spray nozzles with tangential jets, with the spray nozzle mounted on the end of a flexible whip suspended down into the center of the interior. As the cleaning solution sprays out of the jet, the force of the expelled water causes the jet to spin around and the whip to wildly swing back and forth, spraying the cleaning solution randomly all over the interior of the tank.

Automatic Washing Limitations

Because no physical scrubbing occurs with automatic wash systems, the cleanser relies on surfactants and detergents to dissolve the fats left on the interior of the tank by the cream in the milk. However, this is not sufficient to remove milkstone buildup, and the tank may need to be washed occasionally with milkstone remover to remove this scale buildup that can harbor bacteria and contaminants.

Automatic scrubbing only cleans the interior of the tank. It is not capable of cleaning the exterior of the tank, and it does not do a good job of washing around the cover seals. While it is possible to just clean the interior and call it good enough, it does not provide the maximum sanitation of manually washing down the exterior of the tank following or during the automatic wash process. Also, some components that contact the milk such as the drain valve cannot be properly cleaned automatically without disassembling the valve and retaining washer and directly scrubbing in soapy water.

Operating Costs

Substantial reductions in running costs can be made when an ice builder is used in conjunction with off-peak electricity. Pre-cooling milk using a plate or a tube cooler supplied with mains or well water can also reduce costs and add to the cooling capacity of the tank.

Bulk tank condenser units, which are not an integral part of the tank, should be fitted in an adjacent, suitable and well ventilated place.

If at all possible, condenser units should not be fitted on a wall facing the sun. They should be installed in a way which allows them to draw in and discharge adequate quantities of air for efficient operation.

Bulk tank should be easily accessible by large bulk collection tankers and positioned so that the tanker approaches can be kept clean and free from cow traffic at all times.

Although tanks have been calibrated when first installed, bulk tank miscalibration is not uncommon and in some cases it can result in significant loss of income. Milk tanks calibrated on the low side, can cheat raw milk producers by up to 22 litres on each shipment. It is therefore advisable to re-calibrate a bulk tank.

Other usage of Bulk Tanks

Stainless steel bulk tanks are also used to heat or cool a fluid or simply to keep it isolated and warm/cold. Because of the hygienical finishing of the inner and outer side of the tanks, almost any fluid can be stored: water, fruit juices, honey, wine, beer, ink, paint, cosmetics, aromatic food-additives, bacterial cultures, cleansers, oil, or blood.

Automatic Milking

Automatic milking is the milking of dairy animals, especially of dairy cattle, without human labour. Automatic milking systems (AMS), also called voluntary milking systems (VMS), were developed in the late 20th century. They have been commercially available since the early 1990s. The core of such systems that allows complete automation of the milking process is a type of agricultural robot. Automated milking is therefore also called robotic milking. Common systems rely on the use of computers and special herd management software. Also it used to monitor the health status of cows.

AM-systems include single stall systems with integrated robotic and milking functions and multi-stall systems with a transportable robot device, combined with milking and detachment devices at each stall. Single stall systems are able to milk 55-65 cows several times a day, while multi-stall systems with 2 to 4 stalls milk 80 to 150 cows up to three times per day. Automatic milking strongly relies on the cow's motivation to visit the AM-system voluntarily. The main motive for this is the supply of concentrates dispensed in a feed manger in the milking box during the milking process. An automatic milking system has to take over the 'eyes, ears and hands' of the milker. Therefore such a system is equipped with electronic cow identification, cleaning and milking devices and computer controlled sensors to detect abnormalities in milk, in order to meet international legislation and hygiene rules from the dairy industry. Teat cleaning systems include brushes or rollers, inside teat-cup cleaning or a separate 'teat cup like' cleaning device. Several trials showed that cleaning with a device is better than no cleaning, although these systems are not as good as manual cleaning by the herdsman. AM-systems are also equipped with sensors to observe and to control the milking process. Data are automatically stored in a database and the farmer has a management program to control the settings and conditions for cows to be milked. Attention lists and reports are presented to the farmer by screen or printer messages. The AM-system also provides remote notification to the farmer if intervention is required.

Automatic Milking and Management Aspects

Switching from a milking parlor to automatic milking results in big changes for both the herdsman and the cows. Although with AMS immediate supervision of milking is eliminated, new labor tasks appear. They include control and cleaning of the AM-system, twice or three times a day checking of attention lists including visual control of the cows and fetching cows that exceeded maximum milking intervals. Compared with the conventional twice daily milking there is an average of 20% - 30% reduction in total labor, but large variations between farms can be found. Especially in the first transition year, automatic milking might take extra time.

The biggest change without doubt lies in the nature of the labor. The physical work of machine milking, is replaced with management tasks such as frequent checking of attention lists from the computer and appropriate follow up. This work is less time bound than parlor milking, so the input of labor is more flexible. This is attractive on family farms. But because milking is continuous, and system failures can occur anytime there must be a person "on call" at all times. System failures and associated alarms typically occur about once every two weeks although this varies with the level of maintenance and management. In terms of the impact on cows, the AM- system is not suitable for all cows. Poor udder shape and teat position may make attachment difficult and some cows may not be trainable to attend for milking voluntarily. In new installations, the number of cows found to be unsuitable is generally reported to be less than 5%. In the transition from conventional to automatic milking, cows must learn to visit the AM-system at other than traditional milking times. Training and assistance in the first weeks should involve quiet and consistent handling, so the animals adapt quickly to the new surroundings and milking system.

Milking Frequency

In practice, the average number of milkings per cow day varies from 2.5 till over 3.0 milkings per day, but rather big differences in milking intervals are reported by commercial farms. A typical figure is presented in figure. Almost 10% of the cows realized a milking frequency of 2 or lower over a two year period milking with a single stall AM-system. This occurred even though cows with a too long interval were fetched three times per day.

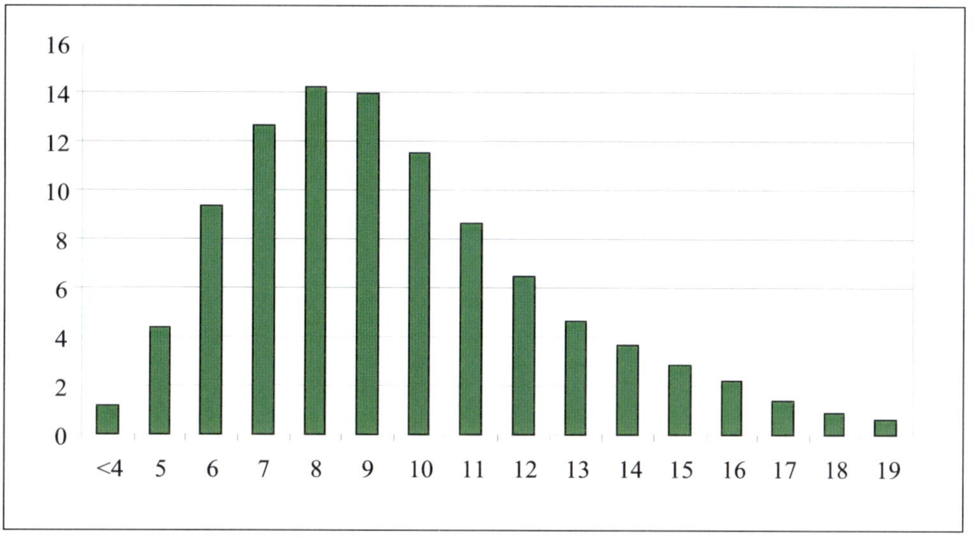

Frequency distribution of milking intervals in hours over a 2-year period.

Such cows will not show an increase in yield or may even show a production loss. By changing the milking parameters of the AM-system, it is quite easy to prevent cows from being milked at low yields or short intervals. But it is much more difficult to prevent cows from being milked with long intervals. This means it will be necessary to manage the intervals by fetching cows that have exceeded a maximum interval. Usually this is done several times per day at fixed times around the cleaning procedures of the AM-system.

Increase in Milk Yield

One of the benefits of automatic milking is increased milk yield from more frequent milking. An increase from 6 to 25% in complete lactations has been shown when milking frequency increases from two times to three times per day. As can be deducted from figure several cows are milked less than twice a day in automatic milking and may therefore produce less milk. Data from all over Europe indicate a production increase of 5-10% for herds milked automatically, however large variations can be seen. In many larger US herds with highly automated conventional parlors, 3 times daily milking is commonplace. For 3x herds adopting automatic milking, a production decrease of 5 to 10% would be expected.

Dynamic Milking and Feeding

Within automatic milking systems, a considerable variation in milking characteristics between individual cows, as well as in interval sensitivity as in feed efficiency can be seen. Current decision support systems do not account for this variation. This is due to lack of applications that estimate individual response parameters from real time process data. Suitable models to enable on-line estimation are dynamic linear models

(DLM), based on a Bayesian approach to time series analysis. A prototype for dynamic milking and feeding was developed and tested on the high-tech farm of Wageningen UR Livestock Research, both for milking frequencies as concentrate feeding. The first results show a considerable increase in feed efficiency and a better utilization of the AMS capacity.

Sensors and Data Handling

AM-systems are also equipped with sensor technology and integrated data management systems to observe and to control the milking process. Data are automatically stored in a database and the farmer has a management program to control the settings and conditions for cows to be milked. Attention lists and reports are presented to the farmer by screen or printer messages. With their sensors AM systems collect enormous amounts of data, which have to be processed with appropriate software. The challenge for both manufacturers as end users is to detect in the data the abnormalities, so actions can be taken. Because abnormalities are rare this is called management by exception. One of the problems for example is clinical mastitis, especially in relation to abnormal milk. By definition, milk of cows that suffer from mastitis has an abnormal visual appearance. It is also one of the most frequently occurring diseases in dairy cattle, and is responsible for the majority of abnormal milk. Despite this, milking a cow with abnormal milk is a rather exceptional event on most dairy farms. As an example, assume that abnormal milk is always caused by mastitis, that 25% of all cows have one case of mastitis each year, and that each mastitis case causes 10 milkings with abnormal milk. For a 100 cow herd with 310 days in milk per cow per year and 2.5 milkings per cow per day, only 0.32% of all milkings will be abnormal. The figure clearly indicates that, even with a high mastitis frequency, the percentage of abnormalities is very low.

Modern AM-systems are equipped with various sensors ranging from sensors to control the milking process till sensors that analyze the milk quality in several ways, like milk composition, cell counts, blood detection, conductivity, progesterone, and so on. All these sensors require smart data handling solutions in order to help the farmer to make the right decision. Special guide lines for automatic milking were developed and approved for automatic milking systems and involved sensor technology within the framework of the International Standards Organization.

Attitude and Expectations

One important factor in successful implementation of an AM-system is the attitude and expectation of the dairy farmer. Almost all manufacturers have had some customers switching back to conventional milking systems. While there is considerable variation in level of satisfaction with different types of systems, an estimated 5-10% of owners have switched back to conventional technology. In some cases expectations were not realistic, in others farmers were unable to adapt to the different management style, and in some cases a high rate of failures of the AM-system discouraged the farmer to continue. However the figure seems to decrease in more recent years, showing that farmers improved their decision process before buying and also that hardware technology has been improved. During the start up period, automatic milking requires a high input of labor and management. Key factors of a successful implementation of AM-systems are:

- Realistic expectations;
- Good support by skilled consultants before, during and after implementation;
- Flexibility and discipline to control the system and the cows;
- Ability to work with computers;
- Much attention to the barn layout and a good functioning cow traffic;
- Good technical functioning of the AM-system and regular maintenance;
- Healthy cows with good feet and 'aggressive' eating behavior.

Barn Layout

Since automatic milking systems largely depend on voluntary attendance, a well laid-out free stall barn is essential for success. The main motive for a cow to visit the AMS is the concentrate provided in the manger of the milking box. Cows should have easy access to the milking stall and selection gates: long alleys, steps and other obstructions should be avoided. A central location for the AM-system minimizes walking distances of the cows. In many countries regulations from the dairy industry require that the AM-system has to be located close to the milking room and that the system is accessible to the operator or the service technician via a clean route.

Cow Traffic

After visiting the milking system, the cow should have access to the feeding area. In "forced traffic" systems she has to pass the milking system in order to get access. In "controlled traffic" systems one-way-gates, with cow identification and selection capabilities, restrict cows to go directly to the feeding area only when the interval since the last milking exceeds the pre-set minimum.

AM-systems with a high occupancy rate and forced traffic may affect the number of visits to the roughage station or feeding fence. In "free cow traffic" systems access to the feeding area is unrestricted and only the concentrate fed in the AM system is used to attract cows. Forced traffic can improve the number of voluntary milkings, however it alters eating and lying behavior. This might even result in reduced intake of roughage and therefore also production losses. Free, semi-free and forced cow traffic have been tested in several studies. Harms observed that free cow traffic resulted in lower milking frequencies and higher number of fetched cows due to long milking intervals. Milk yield did not differed much, but tended to be higher in the free traffic system. Unlimited access to feed is a prerequisite for cow traffic and milking visits. Harms (2004) reported that in all tested traffic systems repeated lack of feed resulted in decreased number of milkings.

There is general consensus that for animal welfare, (semi) free cow traffic is preferred An example of the semi free cow traffic is the Feed-First principle which is used as a mix between controlled cow traffic and free cow traffic. The system also relies on the cows motivation to eat. Without doubt, a well-functioning cow traffic is a key-factor for successful automatic milking. This can be achieved when feed is available for 24h/ day.

Grazing

In most European countries, grazing during summer time is common routine, in some Scandinavian countries like Sweden even compulsory. Moreover, from an ethological point of view, many consumers in North Western Europe believe grazing is essential for cows and several Dutch dairies pay a premium for milk from grazed herds. In the Netherlands grazing is common practice (>80%). The basic principle of grazing cows is that they need to walk at least twice a day to a barn with a milking parlor to be milked. This is faced with problems. The development of large-scale farms and the related large dairy herds lead to long walking distances, increased labor demands and, in many cases, milk production losses. Often the allocation of land makes grazing impossible or leaves too little area for the purpose, causing high leaching potentials due too large urination and defecation surplus.

The introduction of automatic milking induced extra problems, caused by the reluctance of cows to come voluntarily when distances to the barn are longer than 500 meters. In The Netherlands only about 52% of the farms with an AM-system apply grazing, showing on one hand that grazing in combination with AM is less common, but on the other that it is still possible. Overcapacity of the AM-system, e.g. fewer cows per AM-system can partly compensate but this solution is inefficient due the increased fixed costs of the robots. In respect of capacity use of the milking system and percentage of cows to be fetched, restricted grazing systems perform better than unrestricted. Walking distances of up to 500 meters seem to be of little influence on the frequency of robot visits, however longer distances show a negative effect. Moreover the natural synchronized behavior of the cows when moving to the milking unit in the barn causes cueing in front of the AM-system and also ineffective feeding. Several solutions are available to overcome (partly) the disadvantages. Smart pasture gates or selection boxes that control the cow traffic to and from the pasture can be a very helpful tool to the farmer. Only those cows that have recently been milked are sent outside, while the cows that still need to be milked, have to stay inside. Cows have to be attracted to go outside and inside the barn. Grass quality, walking distances, cow health, weather conditions and water supply are important factors within a grazing systems. It is widely accepted that water supply should not be limited, nor be used as a motivation for cow to return to the barn.

Grazing is critical to low cost milk production in Ireland, Australia and New Zealand. An extensive research project in New Zealand, the "Greenfield" project showed that automatic milking in a 100% grazing system under very different circumstances than those found in Europe is quite well possible.

Mobile Automatic Milking Systems

While AM-systems are always positioned in a fixed place within the barn, some years ago the idea was born to design a mobile milking robot. Recently prototypes of mobile automatic milking equipment have been developed in Denmark and the Netherlands. A Mobile Automatic Milking System (MAMS) follows the cows to the pasture. This solves the problem of long distances between the grazing cow and the immobile automatic milk system in the barn. The development of mobile automatic milking systems might also be an opportunity to maintain grazing with bigger herds and on grassland not close to the farm. While the Danish mobile milking system uses a transportable automatic milking system positioned for several months at the pasture, the Dutch mobile milking system can drive from one place to another. The system has caterpillar tracks resulting in a very low pressure per square centimeter to prevent grassland damage. The system is equipped with all necessary machinery bulk tank, vacuum pump, cleaning systems and an automatic milking system so it can operate independently for 2 days. The system was tested in 2008 during the grazing season with a herd of 35 cows with good results, both technically as with respect to the cows. During the 2009 grazing season a herd with 60 cows was managed by the mobile milking system. The research focused on the grazing system and the relation with cow traffic and the milking frequency. The mobile milking system was capable to combine a herd of 60 cows with 24 hour grazing without additional roughage supply while producing a rolling average of 7500 kg milk/year. Offering fresh grass when passing the milking robot really motivates cows to visit the robot. Development of selection units preventing individual cows to enter fresh grass without being milked might or strip grazing might increase milking frequencies.

Milk Quality

Milk quality is a critical concern on modern dairy farms because milk payment systems are based on milk quality and consumers expect a high level of quality and safety from the milk products they buy. Although automatic milking uses the same milking principles as conventional milking, there are major differences. The AM-system is in use for 24 hours continuously. Visual control during the milking process is not possible. Cows will visit the AM-system more or less voluntarily and this will result in a big variation in the milking frequency from cow to cow. All these aspects may influence the quality of the milk produced.

Somatic cell count (SCC) and Total Bacterial Count (TBC) are, respectively, measurements of the number of white blood cells and the total number of bacteria present in a milk sample. A high SCC might indicate reduced udder health due to Mastitis (udder infection) and implies a lowered milk quality. The cleaning of the milking equipment and the cooling of the milk seem to be the most important factors regarding the increase in TBC. In general AMS herds consistently show slightly higher SCC and TBC values than conventionally milked herds. Table presents the results for 4 quality parameters. In general differences are relatively small and far within the requirements of the dairy industry. Nevertheless with 24h operation, milk hygiene requires continuous attention from the herdsman. Special attention has to be paid to teat cleaning and the cleaning of the teat cups and the milking machine including transfer line to the bulk tank, Hygiene management should not only focus to the automatic milking system, but also to the general hygiene standards in cubicles and the floors in the barn.

Table: Milk quality results for farms before and after introduction of AM-system.

	Conventional milking		Automatic Milking	
	Two times milking	Three times milking	Before	After
Bacterial count (1000/ml)	8	8	8	12
Cell count (1000/ ml)	181	175	175	190
Freezing point (°C)	-0,520	-0,521	-0,521	-0,516
Free fatty acids (meq/100 g fat)	0,44	0,54	0,41	0,59

Free Fatty Acids

Free Fatty Acids (FFA) in milk are formed as a result of hydrolysis of milk fat by milk lipase, also called lipolysis. Milk fat is protected from lipase activity because the milk fat is protected by a membrane around the milk fat globules. Lipolysis can be initiated either by physical damage to milk fat globules during milking or pumping, stirring or cooling of milk or by animal-related factors such as the cow's health status, lactation stage, the frequency of milking and dietary factors.

It is generally known that the content of free fatty acids (FFA) in milk will increase with shorter milking intervals, all the more if the yield per milking is rather low. All studies with AM-systems show a significant increase in FFA levels for AM-systems. This increase cannot be explained solely by the shorter intervals, because the increase of FFA with AM-systems is even bigger than with conventional milking parlors milking three times per day. In a Danish study small but statistically significant differences in average FFA concentration of milk between five different brands were detected. Apart from differences in diameter or length of "short" milk tubes, the technical specifications of the AM systems could not explain these differences.

AM systems generally have about 3 to 4 times higher air intakes (air inlet) during milking than conventional milking clusters. It is known that excessive air intake increases lipolysis in raw milk. The higher air intake of AM systems has been recognized as a factor that can partially explain higher FFA concentrations of raw milk. Rasmussen et al. also reported on visits to 55 farms with high FFA levels. For 31 conventional farms most frequent faults were air leakage (71% of the herds) and intake of too much air in the cluster (61%) and pumping and stirring faults occurred on 29% of the farms. The main faults for 24 AM system farms concerned stirring of milk at very low levels in the bulk cool tank (79%), pumping of milk (67%) and the cooling of milk (58%). Low milk yields and low-energy diets increase the FFA level of milk. However, many aspects of the role of nutritional factors in lipolysis and FFA levels are still unclear and require further investigations.

Animal Health

Within a large EU project Automatic Milking carried out between 2000 and 2004, special attention was paid to animal health. In Denmark, The Netherlands, and the UK, 15 herds each were selected for monitoring the impact of transition to automated milking on animal health. The herds recruited represented the types of AMS marketed in each country. Each farm was visited at least twice before installation of the AMS and a minimum of twice, but often up to six times, after installation. On these visits assessments were made of at least half of the cows or fifty animals on body condition and locomotion, and forty cows for teat condition (on some farms in the Netherlands and UK only). Farm data including milk production, milk quality, animal records on individual cow milk cell count, fertility, animal treatments, animal movements, veterinary purchases were collected.

The body conditions varied more between countries than in response to the introduction of AM.

No change in locomotion was seen one month after AM installation. The scores in Denmark and UK increased slightly by 3 months after installation, but not significant. In the UK the average score increased on seven farms whilst unchanged on 6 farms. Scoring was continued on 12 of the UK farms. Twelve months after installation of AMS the lameness has increased significantly. Prior to installation eleven of fourteen UK herds were grazed but only six after installation. The poorer locomotion may reflect the increase in constant housing. The overall impact of conversion to AM was assessed by comparing how each individual farm handled the main indicators of animal health during and after the transition to automatic milking. Comparing 12 Dutch farms only one farm improved in locomotion, body condition as well as cell counts. Overall, little change was apparent. Locomotion improved in five herds and deteriorated in five herds. Body condition score decreased in eight herds but only by a small amount. It increased in two herds but not making the cows any fatter, just more typical. The only major deterioration found was an increase in somatic cell count and the proportion of cows with a cell count above a threshold, where only two of the herds produced better quality milk. Overall there was little evidence of major changes occurring in the common measures of fertility. None of the changes were statistically significant but all suggestive of poorer fertility, at least in the transition period from conventional milking to AM.

Hillerton et al. conclude that no major problems in converting from conventional milking to AM could be identified but equally none of the 44 farms has been found to achieve a substantial improvement in any aspect of cow health. The transition period to AMS comprises a period of higher risk to health that extends from weeks before installation when resources start to be diverted from cow management. The length of the transition will vary on individual farms related to many unique factors. Several potential problems may develop in the longer term and anticipation of these is necessary.

A large study to risk factors related with automatic milking in The Netherlands showed that risk factors for mastitis are more or less comparable with those found at conventional milking. However the udder health status in the period just before the transition to automatic milking showed a large correlation with the udder health status afterwards. Other factors are mainly related to management strategies and the use of information and data from the AM-system.

Economical Aspects

Investment required for AM-systems are much higher than for conventional milking systems and thus the fixed costs of milking are higher. However more milk with less labour means that the costs of milking per kg of milk will decrease. Theoretically, with an AM-system more cows can be kept with the same labor force than with conventional milking, but this may involve additional investments in buildings, land or feed and perhaps in milk quota. On a farm with more than one full time worker the possibility exists to reduce labor input and thus costs. Quite often that does not happen and the time saved as a result of lower labor requirement is used for personal activities. Meskens and Mathijs (2002) found that two third of AM-farmers state social reasons for investing in automatic milking, such as increased labor flexibility, improved social life and health concerns. In parts of North America, with large-size herds and numerous milkers, it may turn out that savings on labor costs may become a decisive motive to implement automatic milking.

Simulation Models

Several simulation models have been developed to calculate the pure economic effect of investment in automated milking. The “Room for Investment” model computes the amount of money that can be invested in an AMS, without a decrease in net return compared with conventional milking. The RFI-value calculates the annual accumulated return from increased milk yield, savings in labor, and savings in not investing in a milking parlor and divides this by the annual costs of the AM-system. The model can use farm specific factors and circumstances to calculate the RFI-value. Figure shows the results of a combined sensitivity analysis illustrating that increased milk yield and labor savings are essential factors regarding the economy of automatic milking. The RFI-value for the basic farm (700.000 kg milk, 8500 kg milk per cow per lactation, 82 cows, 75 hrs of labor per week) with 500 kg per cow yield increase, 0,75 hour net labor saving per day (~10% labor saving), compared with a automated milking parlor and 20% annual costs of the AM-system amounts € 137.000. In this example this would be sufficient for investment in a single box (investment ~ €120,000). Both labor saving and yield increase have a large effect on the RFI value.

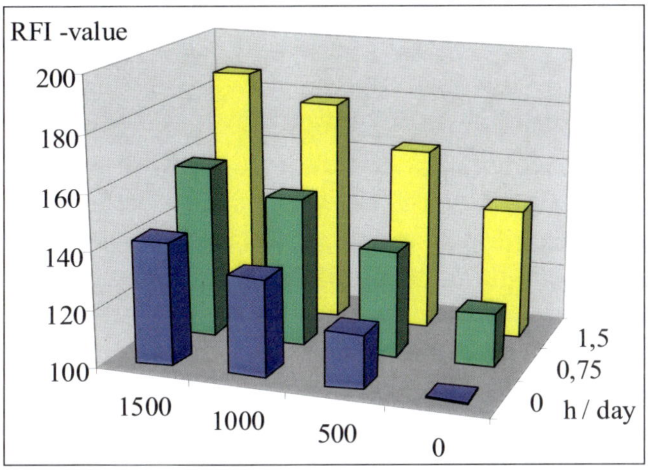

Increase in yield (kg) per cow. Room for Investment (RFI) due to labor saving and milk yield increase with annual costs for AM-system of 25% of investment. Comparison made with an highly automated milking parlour.

American models suggest that on large farms, as the hours of use of large automated milking parlors increase, their relative economic advantage over AM-systems increases. These studies showed that on farms with 60 to 180 cows, capital investment in automatic milking is only slightly higher than in conventional milking systems, in part because of smaller space requirements than for parlors with large holding areas. Lower labor input for AM- systems makes them competitive in this herd size range. For herds with more than 240 – 270 cows, extending the hours of use of conventional parlors, without additional capital investment made parlor milking much more efficient. It follows that in the central and western United States, where herds of 500 to several thousand cows predominate, widespread adoption of automatic milking will not occur unless capital cost of these systems decrease or labor costs increase substantially. Since both of these parameters are highly likely to trend in this direction, more widespread adoption of automatic milking in nearly all areas of the developed world would appear to be only a matter of time.

Economic Results from Commercial Farms

Economic results from commercial farms show that the use of AM-systems in general is profitable, although highly depending on the extra milk yield and labor savings. A Dutch case control study between farms with AM-systems and farms with conventional milking systems showed no differences in margin, although fixed costs for the AM-farms were higher. AM-farms saved 29% labor and therefore when economical results were transformed to full time equivalents (FTE), AM-farms in the case control study had greater revenues, margins, and gross margins per FTE than the farms with conventional milking systems. So when deciding between investment in an AM-system or in a conventional milking system, dairy farmers must weigh decreased labor needs for the AM-system against the increased fixed costs of milking with an AM-system. Therefore in many cases adoption of an AM-system is for many dairy farmers a socio-economic decision, rather than just a purely economic decision.

The number of farms milking with automatic milking has increased significantly since 2000. In areas where labor is expensive or in short supply automatic milking is a valid alternative to traditional parlor milking. However if labor is available, and particularly where herd sizes are large conventional milking, often with rotary or rapid exit parlors equipped with features to increase throughput per man hour will remain popular. The introduction of automatic milking has a large impact on the farm and affects all aspects of dairy farming. Because milking is voluntarily there is large variation in milking intervals. Both farm management and the lifestyle of the farmer is altered by automatic milking. AM-systems require a higher investment than conventional milking systems but increased milk yields and reduced labor may lead to lower fixed costs per kg milk and increased margins per FTE.

Successful adoption of automatic milking depends on the management skills of the farmer and the barn layout and farming conditions. Animal health and well-being is not negatively affected by automatic milking, but till now no particular benefits for the health of the cows have been found. Grazing is a common routine in many countries and although grazing with automatic milking requires extra attention from the herdsman, grazing is still profitable. New solutions like mobile milking systems might be an alternative for large herds. Improved sensor technology will help farmers to manage the individual cow within the herd. A better understanding of the characteristics of automatic milking systems will help farmers to make the right decision. Both conventional and automatic milking will be used on dairy farms in modern dairy countries in the foreseeable future.

PERMISSIONS

All chapters in this book are published with permission under the Creative Commons Attribution Share Alike License or equivalent. Every chapter published in this book has been scrutinized by our experts. Their significance has been extensively debated. The topics covered herein carry significant information for a comprehensive understanding. They may even be implemented as practical applications or may be referred to as a beginning point for further studies.

We would like to thank the editorial team for lending their expertise to make the book truly unique. They have played a crucial role in the development of this book. Without their invaluable contributions this book wouldn't have been possible. They have made vital efforts to compile up to date information on the varied aspects of this subject to make this book a valuable addition to the collection of many professionals and students.

This book was conceptualized with the vision of imparting up-to-date and integrated information in this field. To ensure the same, a matchless editorial board was set up. Every individual on the board went through rigorous rounds of assessment to prove their worth. After which they invested a large part of their time researching and compiling the most relevant data for our readers.

The editorial board has been involved in producing this book since its inception. They have spent rigorous hours researching and exploring the diverse topics which have resulted in the successful publishing of this book. They have passed on their knowledge of decades through this book. To expedite this challenging task, the publisher supported the team at every step. A small team of assistant editors was also appointed to further simplify the editing procedure and attain best results for the readers.

Apart from the editorial board, the designing team has also invested a significant amount of their time in understanding the subject and creating the most relevant covers. They scrutinized every image to scout for the most suitable representation of the subject and create an appropriate cover for the book.

The publishing team has been an ardent support to the editorial, designing and production team. Their endless efforts to recruit the best for this project, has resulted in the accomplishment of this book. They are a veteran in the field of academics and their pool of knowledge is as vast as their experience in printing. Their expertise and guidance has proved useful at every step. Their uncompromising quality standards have made this book an exceptional effort. Their encouragement from time to time has been an inspiration for everyone.

The publisher and the editorial board hope that this book will prove to be a valuable piece of knowledge for students, practitioners and scholars across the globe.

INDEX